THE BEAUTY OF FALLING

More praise for
THE BEAUTY OF FALLING

"The Beauty of Falling is a great book for young women in STEM to check out not only for interesting perspectives on gravity, but also to connect with de Rham through her story of being a woman in science."
—HAILEY DICKINSON, *Yahoo!*

"[De Rham's] sense of wonderment is infectious."
—TONY MIKSANEK, *Booklist*

"Those who enjoy the work of James Gleick and Brian Greene will find [The Beauty of Falling] an intriguing read. At the leading edge of science, this book combines innovative research with a personal story."
—*KIRKUS*

"[De Rham] provides an entertaining and insightful history and explanation of gravity infused with personal reminiscences and anecdotes."
—*CHOICE*

"The Beauty of Falling is an enjoyable and relatively quick read: a first-hand and personal glimpse into the life of a theoretical physicist and the process of discovery."
—KURT HINTERBICHLER,
CERN Courier

"A unique and very personal story about the joy of theoretical physics, taking the reader from the fundamentals to the cutting edge—and bubbling with joy and wonder throughout. Highly recommended."
—CHRIS LINTOTT, astrophysics professor, author,
and presenter on the BBC's *The Sky at Night*

"A wonderfully clear, right up-to-date account of gravity and what we have learned about it and from it."
—JOCELYN BELL BURNELL, visiting professor,
University of Oxford

"In this engaging firsthand account, de Rham brings the story of gravity to life, sharing her deep insights into its fundamental nature. From her highly expert vantage point, she elegantly intersperses the fascinating science with her own life story and her quest to make sense of gravity's mysteries."
—JO DUNKLEY, author of *Our Universe*

THE BEAUTY OF FALLING

A Life in Pursuit of Gravity

Claudia de Rham

PRINCETON UNIVERSITY PRESS

PRINCETON & OXFORD

Published by Princeton University Press
41 William Street, Princeton, New Jersey 08540
99 Banbury Road, Oxford OX2 6JX

press.princeton.edu

GPSR Authorized Representative: Easy Access System Europe -
Mustamäe tee 50, 10621 Tallinn, Estonia, gpsr.requests@easproject.com

All Rights Reserved

First paperback printing, 2026
Paperback ISBN 9780691237503

The Library of Congress has cataloged the cloth edition of this book as follows:

Names: Rham, Claudia de, author.
Title: The beauty of falling : a life in pursuit of gravity / Claudia de Rham.
Description: Princeton, New Jersey : Princeton University Press, [2024] | Includes bibliographical references and index.
Identifiers: LCCN 2023022642 (print) | LCCN 2023022643 (ebook) | ISBN 9780691237480 (hardback) | ISBN 9780691237497 (ebook)
Subjects: LCSH: Gravitation—Popular works. | Gravitation—History—Popular works. | Relativity—Popular works. | Relativity—History—Popular works. | Gravitational waves—Popular works. | Rham, Claudia de.
Classification: LCC QC178 .R55 2024 (print) | LCC QC178 (ebook) | DDC 531/.14—dc23/eng/20231012
LC record available at https://lccn.loc.gov/2023022642
LC ebook record available at https://lccn.loc.gov/2023022643

British Library Cataloging-in-Publication Data is available

Editorial: Ingrid Gnerlich, Whitney Rauenhorst
Production: Jacqueline Poirier
Publicity: Maria Whelan (US), Kate Farquhar-Thomson (UK)

Cover design by Karl Spurzem

This book has been composed in Arno Pro with Futura New

To my little trolls

Contents

A Journey through Gravity

Gravity. Such a familiar concept, present in every language and culture, yet one that scientists have struggled to understand for millennia. It is the overarching miracle connecting everything, everywhere, forever in the Universe. Universal in every sense. As humans, we may think of it as the hidden force that keeps us firmly planted on Earth, the reason why the Earth orbits the Sun, or the interaction that allowed the Milky Way and its hundreds of billions of stars to form. But that barely hints at its true significance. Gravity is the reason why the Universe itself can even exist and evolve. It elevates space and time from mere pieces of scenery into central actors in the unfolding drama of reality. As we embrace gravity, we can't help but also pit ourselves against it: leaping, floating, or flying as we pursue brief moments of freedom from its command. I, for one, have been chasing gravity my entire life—seeking, like so many scientists who have come before me, to unravel its deepest mysteries.

Imagine yourself alone in the cockpit of a small, single-engine aircraft, patiently waiting on the taxiway for the signal from air traffic control. Four simple words, "clear to take off," resonate like a magical password, unlocking a precise series of events that will achieve what would have been impossible just over 120 years ago: lifting and floating a one-ton object into midair. As the craft zooms down the runway, you are pressed back into the foam of your seat, accelerating horizontally to a speed of 100 to 200 km/h. Ironically, it is this horizontal speed that will allow the pressure under your wings to overcome the vertical pull of gravity and lift you skyward. As you rise to cruise altitude, even the slightest amount of turbulence shakes the small plane. For a moment, you feel like you are trapped inside a snow globe, existing at the mercy of some titanic, mischievous shaker—until you remember that a little tweak of the rudder, a gentle push on the trims, or a subtle twist of your ailerons is all you need to take control and surf gracefully on the airflow, simultaneously pushed upward by the pressure and pulled downward by gravity.

If soaring above the clouds is not for you, perhaps you would prefer to picture yourself submerged in the deep blue, mingling with thousands of coral reef fish a few dozen meters below the surface. As you contemplate the serenity of this underwater world, you are plunged into a silence broken only by the popping sounds of the vibrant coral reef and that of your own breath as you slowly inhale from your air tank and exhale small bubbles that shoot to the surface. With each breath, your body gently bobs up and down a few centimeters as the pressure of the air in your lungs tries to compensate for the force of gravity and the mass of the column of water that is pressing against every cell in your body.

Flying high in the air and diving deep under the sea are two of the most thrilling ways to defy gravity, at least here on Earth. But to achieve the ultimate feeling of weightlessness, nothing compares to floating in space, seemingly escaping gravity's clutches altogether. The feeling of freedom is no longer an illusion—there are no strings to pull or pressures to counteract. Observing Earth from orbit, you can savor the absolute freedom of free fall, a concept deeply engrained in our understanding of gravity, even while it remains a luxury that few have had the opportunity to enjoy.

In my life, I have experienced the joy of flying and diving and came within a hair's breadth of making it to outer space. But we don't need a fancy plane, scuba gear, or space shuttle to experiment with gravity. In fact, whether we are doing something as simple as dropping a ball, swinging in a hammock, or skipping a stone, we're all scientists conducting our own personal experiments and drawing our own conclusions about this universal yet mysterious phenomenon.

But what exactly is going on in those moments? What is gravity? It seems like such an innocent question, yet the answer always seems to be hidden behind abstruse laws of physics. Physical phenomena are often portrayed as a set of obscure fundamental rules—Archimedes' principle, Newton's inverse square law, Bernoulli's principle, and the like—that nature must unquestionably and rigidly obey. These laws are, of course, central to our understanding of the world and the structure of our reality. They have revealed how buoyancy allows boats to float, and how the difference in pressure caused by the motion of the air beneath their wings allows birds and planes to navigate the skies. They have enabled us to send a man, and hopefully soon a woman, to the Moon. Yet the presentation of these laws as being set in stone belies our scientific history. Far from being immutable and unchanging, our understanding and appreciation of these laws—what they mean, where they come from, and what lies behind them—is continuously unfolding before us.

Galileo Galilei, Johannes Kepler, Sir Isaac Newton, Albert Einstein, Stephen Hawking, Sir Roger Penrose, Andrea Ghez, and countless other brilliant scientists have each brought a new perspective to our understanding of gravity, but our journey is far from finished. Think of this book, then, as an invitation to join me in the quest to uncover the meaning of gravity, to grasp its connection with the structure of reality. Fortunately, for the most part we will not be undertaking this adventure alone. Instead, we will be guided by some of the greatest scientific minds of the past several centuries—that is, at least, until we reach the edge of the map, where we will take some exploratory steps into the unknown. Our journey will begin, however, in well-charted territory with a few trustworthy companions.

With their realization that gravity must be a universal force, acting on everything and accelerating everyone in the same way, regardless of their mass, Galileo, Kepler, and Newton provided the first crucial piece of the puzzle. This insight was made possible by a new perspective on what it means to *be free*, a perspective that discarded centuries of Aristotelian dogma and radically transformed the concept of *inertia*.

This new perspective was brought to light in 1632, with the publication of Italian astronomer and physicist Galileo Galilei's *Dialogue Concerning the Two Chief World Systems* (*Dialogo sopra i due massimi sistemi del mondo*). In the dialogue, Galileo championed a new Copernican revolution, one that went beyond merely denying that the Earth occupied a special place in the solar system, by further dismissing the idea that *any* person or object could ever hold a privileged position with respect to the laws of nature.

To make this argument, Galileo considered the world through the eyes of a sailor confined to the main cabin below the decks of a moving ship. Unable to see the world outside, the sailor was entertained by watching the motion of "some flies and butterflies" with whom she shared the cabin. Galileo realized that the sailor would not be able to tell whether the ship was at rest or in motion at constant speed, at least not from observing these small flying animals. Why? Because if the ship moves at constant speed, so does everything on board, including the air in which the flies and butterflies flutter about. The sailor, trapped below deck, can only observe the motion of the flying creatures relative to the inside of the ship's cabin. Galileo used this thought experiment, which highlighted the importance of *relative* motion, to explain how the Earth could rotate without us being able to feel it. Moreover, once we recognize that we cannot tell the difference between the lower deck of a ship at rest and that of one in uniform motion, we can infer that the laws of physics should be the same for any observer moving at constant velocity, no matter the speed.

It is precisely this notion of "Galilean relativity"—the realization that the laws of nature are the same regardless of who describes them—that is enshrined in Newton's first law of motion, which holds that every object will remain at rest or in uniform motion in a straight line unless compelled to change its state by the action of an external force.[1] Newton realized that *being free* is the privilege to carry on undisturbed, pursuing our journey at the same velocity, uniformly. Building on the work

---

1. This idea replaced the Aristotelian notion of *inertia*—the desire to slow down and come to a state of absolute rest. In contrast to Aristotle, who thought that forces were necessary to maintain velocity, Newton realized that *forces* lead to *acceleration* (change in velocity). In our everyday lives, friction with the air and the ground acts as a force, naturally slowing us down (deceleration or negative acceleration). However, in outer space, where there is no air and no friction, objects can be free and maintain a uniform velocity.

of Kepler, who developed the laws of planetary motion, this insight would later lead to Newton's 1687 law of universal gravitation, also known as Newton's inverse square law. According to this law, the force of gravity exerted between any two massive particles (that is, particles having mass) is a *universal* and *instantaneous* force, whose intensity decays as the square of the distance between the two particles.

Newton's law, as many of us have been taught, describes how an object, when dropped, is inexorably attracted by the mass of the Earth. But the universal nature of gravity extends far beyond this simple phenomenon. It applies to everything and everyone, no matter the object, no matter the separation. In 1798, Henry Cavendish was among the first to test it formally in a laboratory, and more than three centuries after its discovery, Newton's inverse square law has been scrutinized with impeccable precision, from distances smaller than a tenth of the width of human hair to separations that extend billions of kilometers. In fact, Newton's law of universal gravitation is so fundamental that it can still be used to predict how gravity has governed most of the evolution of our Universe, from the gravitational collapse of dark matter to the formation of clusters of galaxies and the creation of the solar system.

Centuries passed before observational evidence began to cast a sliver of doubt on Newton's law of gravity. However, in retrospect, the idea that gravitational attraction between any two objects happens *instantaneously* should have raised a red flag. According to Newton's simple law, if two particles were to appear, they would be *immediately* attracted to one another without any delay. No matter what your views on attraction may be, we all know that this phenomenon cannot be immediate. Even when it comes to love at first sight, you first need to "see" the other person (that is, to "communicate," even if not verbally) for attraction to take place. Newton himself, in a letter to Richard Bentley, expressed his discomfort with the concept of an instantaneous law: "Tis unconceivable that inanimate brute matter should (without the mediation of something else which is not material) operate upon & affect other matter without mutual contact; as it must if gravitation in the sense of Epicurus be essential & inherent in it. And this is the reason why I desired you would not ascribe {innate} gravity to me" [1].

Our own journey will begin two centuries later, when American scientists Albert Michelson and Edward Morley revealed the results of their infamous "failed experiment," ushering in a new scientific revolution. Shortly thereafter, Einstein introduced new ideas of relativity into our understanding of gravity: first putting forward the notion of special relativity, which supplanted the kinematics of Galileo, and then unveiling gravity as we understand it today through the theory of general relativity. Guided by these theories, we will uncover an entirely new structure of physics and understanding of our Universe in which gravity is fundamentally identified with the very fabric of space and time, entwined and unified.

Today, more than a century has passed since Einstein's breakthroughs, and general relativity stands stronger than ever. Gravity has been exhaustively tested, including in some of the most extreme environments, and the evidence unfailingly accords with Einstein's predictions. The very force within gravity has been detected thanks to gravitational waves. At the same time, we have also learned much more about the quantum nature of our world through atomic, nuclear, and particle physics, quantum chemistry, and the numerous technological advances of the electronic and computer age. With these advances, new ideas and theories constantly bubble up in our effort to make sense of the world in which we live. And yet, to date, none has surpassed Einstein's theory of general relativity, despite the obvious need for new physics. For there is one thing that, from the very beginning, general relativity itself has been forthright about: there is a point where the theory must fail, where a brand-new layer of physics waits to be unveiled. From this failure comes the opportunity to probe and appreciate nature on an even deeper level.

As we continue on our journey, we shall see how gravity, viewed from a more modern perspective, can also be thought of as the manifestation of a fundamental particle—the graviton—much like electromagnetism is the manifestation of the photon, the fundamental particle of light. In the very same way that we can "see" *light* as electromagnetic waves propagate through space and time, we can now "hear" gravitational waves (or *glight*, as we shall call it here) as they disturb the very fabric of spacetime. We have now observed gravitational waves through many different instruments, and the reality of glight has become unquestionable. Their detection offers an unparalleled opportunity to decipher the many mysteries that our Universe is still hiding. What is the origin of the Universe? What are the dark components of the Universe that explain its structure and evolution but cannot be directly detected with our instruments? What is our fate? These profound questions are begging for answers. And who wouldn't want to follow that trail?

Eventually our journey will take us to the edge of the map. While Einstein's theory of general relativity has provided natural and elegant answers to some of the most perplexing questions about the nature of gravity, it also has raised several puzzles with which we continue to grapple. How is it that the contributions of known particles that we understand so well in our underground particle accelerators affect the Universe in ways we cannot even start to comprehend?

As we attempt to reconcile the evolution of our Universe with the fundamental quantum nature of the world, we will be forced to reconsider gravity on an even deeper level. What if, on large cosmological scales, gravity behaves differently than predicted by general relativity? What if gravity, long assumed to be massless, in fact has mass? This idea is almost as old as general relativity itself and has been explored by some of the greatest scientists of the past century. Until recently, all attempts to make sense of this idea have failed dramatically. Yet far from being the end, this is where the most exciting part of our journey will begin as I guide you through new pathways that my colleagues and I have recently uncovered in our quest to grapple with gravity.

These paths previously looked so unpromising that their exploration was considered not only impractical or dangerous but simply unthinkable. Today, however, it seems that they may lead us to an entirely new way to think about gravity. And while these new theories may not provide final answers to all of our questions, by exploring gravity as it might be, even if not in our own reality, we may come to appreciate nature for all that it has to offer.

Gravity is one of the first physical phenomena of which we are aware, and we possess a near universal desire to probe its limits. As babies, we repeatedly push toys off the table, watching them tumble to the ground (and watching our exasperated parents retrieve them). As children, we jump tirelessly on the trampoline, seeing how high we can soar before being pulled back down to our terrestrial home. As old friends, we skip stones at the beach, observing the beauty of the cascading ripples. In every instance, we both play with and try to counteract this tenacious phenomenon. Its constant pull is the source of so much stress in our lives, but rather than hiding from it we all learn to embrace it.

As we fall through the curvature of spacetime as freely as we fall through our lives, we soon realize that, while being free and straight, our journey through space and time is far from straight-forward. Certainly, our journey would not be complete without its share of obstacles and falls. Embracing them and appreciating the beauty of falling is essential if we are to make progress in our never-ending quest. All theories of gravity developed so far have experienced the virtue of failure. Daring to fail means appreciating each fall not as an embarrassing epilogue but rather as an opportunity for a more fundamental understanding of nature.

Think of this journey, then, as a celebration of gravity's mysteries and of science itself—complete with its doubts and failures, yes, but also with the incredible thrill of discovery. This is not just my quest, nor that of my colleagues. It is not the discovery of Einstein or Newton alone. It is our *shared* adventure, yours as much as that of the great scientists who paved the way. It is a journey that began thousands of years ago, and one that may never end. Along the way, however, we hope to gain knowledge that will enrich the lives of future generations and civilizations, allowing them to pursue their own destiny, to surf between new layers of reality, and to interact with the all-encompassing fabric of the Universe.

CHAPTER 1

A Universal Language

Although my mother tongue is French, I learned the language from a Swedish mother while living in various Quechuan provinces of a Spanish-speaking country. In retrospect, it is perhaps not entirely surprising that I developed one or two weaknesses and an unidentifiable accent in any language I attempt to speak. When I attended the École Polytechnique in Paris in my early twenties, people would often compliment me on how quickly I had picked up the language, a stinging dismissal of my two decades of experience. A few years later, as I moved to Canada, Québécois would kindly switch to English when talking to me, assuming (quite rightly) that I did not quite feel at home with their language. Ironically, the English-speaking Canadians would do exactly the opposite. And when piloting the tiny Diamond DA20 Katana in which I learned to fly, the air traffic controllers would respond with disbelief when I would make a request to land on the "highway" rather than the runway.

The truth is that I have always struggled to find the right words to express myself to others. When I was two years old, we moved as a family to Ayacucho, a beautiful part of the Peruvian Andes. This was a magical place to spend one's childhood; the colors of the market and the constant music still resonate in my mind.

But Ayacucho was also the base for Sendero Luminoso (Shining Path), a far-left terrorist organization. One afternoon, as we were celebrating a friend's birthday, the festivities were suddenly interrupted by gunshots in the surrounding hills. As my family hurried home, we were stopped at a military checkpoint where one of the soldiers pointed his weapon at my father's ear. Taken over by panic, I started shouting repeatedly "*Dis Pare!*" (French for "Tell him" and Spanish for "Stop!"). To my family's dismay, I had forgotten that "*Dispare!*" was also Spanish for "Shoot!" Fortunately, the soldiers paid little attention to my muddled babbling.

While my translation errors have proven relatively harmless, if not outright comical, I have always been eager to connect with a more reliable, universal language—to make sense of the world in a way that transcends words and misunderstandings. As a child, I took comfort in simple rules, which always seemed to be much clearer to me than the words I struggled to find. For better or worse, simple rules seemed to occur everywhere I looked, including when observing the indefensible activities of the Sendero Luminoso.

Whether it was breaking into the local prison to release key prisoners, attacking government buildings, or kidnapping civilians for ransom, the Sendero Luminoso's tactics remained consistent. Right before they attacked anything or anyone threatening their ideology, their members would blow up a high-voltage pylon after sunset, plunging the entire city into darkness. Then the rule was clear: I would count to ten, and everything would go wild. I would hear shouting, see shooting, and feel the fear and pain everywhere I looked. But the ten-second countdown is what I remember most vividly from those episodes: no panic in that interval, just a simple rule that never failed.

Whatever the circumstances in which we find ourselves, we all have an innate instinct to make sense of the world around us, to derive laws that explain our observations and allow us to make predictions. Doing so allows us to comprehend the previously incomprehensible, which explains, in part, the success of our species. In my case, recognizing these simple patterns and the way they structure our world slowly went from modeling the sadly inexplicable to awakening an awe for the unknown, introducing me to the thrill of discovery and to some of the happiest realizations as they guided me toward wanting to uncover the mysteries of our Universe.

As a five-year-old, swinging in a hammock just outside the city of Iquitos at the border with the Peruvian Amazon, I distinctly recall experiencing the sensation of complete weightlessness. As I gazed at the multitude of stars that pierced through the thousand-year-old trees, I could almost imagine floating in outer space, out of time, and conquering gravity. This moment sparked what would become a lifelong fascination with the subject. Today, my job as a scientist is to strive for a more profound and universal understanding of nature, to investigate the fundamental principles that govern the Universe. Though such scientific investigations may appear to be governed solely by mathematical theorems and physical laws, it is the passion and curiosity that we each possess that give rise to our most profound discoveries.

We are all scientists at heart, recognizing patterns, deciphering their meanings, and predicting outcomes in ways that transcend any form of communication. Perhaps this is why we have been fortunate enough to recognize the elegant universality of gravity and nature as a whole. Or perhaps it is nature's universality that has gifted us with the ability to see the world in such a symmetric and elegant way. Whatever the case, it was our attempt to make sense of nature's patterns that revealed one of the most significant breakthroughs in our understanding of reality.

The Universality of the Speed of Light

From ancient Chinese astronomy born over three thousand years ago to the most recent discoveries of the James Webb Space Telescope, almost every observation we have made, every hint we have gathered about the way nature works, has ultimately been seen through light. Light has served as a valuable, infallible messenger, sharing with us the secrets of our Universe, whether through our own eyes or through telescopes, observatories, or other experiments. Yet only in the past few centuries have we begun to better understand what light really is.

After a series of breakthroughs, in 1861 the Scottish mathematical physicist James Clerk Maxwell summed up thousands of years of wisdom in a remarkably simple set of equations. "Maxwell's equations," as they've come to be known, described everything we knew at the time about the electric and magnetic forces, unifying them into one "electromagnetic field." Four years later, Maxwell deduced from these same equations that disturbances in the electromagnetic field behave and travel like waves. Their speed was calculated to be "so nearly that of light" that Maxwell was led to the inevitable conclusion that the light we see is nothing other than electromagnetic waves, confirming a conjecture made in 1846 by Michael Faraday [2].

In many ways, light acts like ocean waves moving across the surface of our seas or sound waves moving through the air. In fact, visible light, radio waves, microwaves, X-rays, gamma rays, and ultraviolet and infrared radiation are all just the same thing: light. The only difference is the wavelength of the wave, the distance between the peaks of oscillation. Just as music is made from playing notes of different pitches (wavelengths), different colors are produced by light with longer and shorter wavelengths. For instance, red light has a longer wavelength than blue light.

The term "light" is sometimes used to refer exclusively to the visible part of the spectrum—electromagnetic waves with frequencies or wavelengths (roughly between 400 and 700 nm) to which our human eyes are attuned. But all types of electromagnetic waves, regardless of wavelength, are fundamentally the same. Electromagnetic waves could, in principle, have wavelengths as short as the Planck length, or 10^{-35} m—that is, ten billion billion billion times shorter than visible light, well beyond the ultraviolet range. At the other end of the spectrum, far into the infrared range, radio waves with wavelengths as long as a few hundred thousand km can be detected on Earth. If we used the whole Universe as our detector, we could in principle detect electromagnetic waves with wavelengths as large as the observable Universe, or roughly a million billion billion km long. For our purposes, we shall make no distinction between waves at visible frequencies and others in the infrared or the ultraviolet spectrums, and we will use the word "light" to refer to any electromagnetic wave, regardless of its frequency.

Light, like sound, takes some (though much less) time to travel. In Maxwell's theory, light waves move at a fixed speed through empty space. At the time of Maxwell's work, all known waves traveled through a medium. Ocean waves, for example, are oscillations in the fabric of water, while sound waves travel along many different mediums, such as the string of a tin can telephone toy. Without any medium to carry sound, there will be complete silence, giving rise to the ominous warning of the film *Alien*: "In space no one can hear you scream." Maxwell naturally assumed that light must travel through a medium as well. The proposed medium for light was called the "luminifer-ous æther," a substance that pervades the entire Universe. How-ever, the search for signs of this luminiferous æther required a high level of experimental precision, and the initial attempts to detect it produced only questionable successes. Eventually, however, these efforts would culminate in one of science's most famous failed experiments: the Michelson–Morley experiment, which we will discuss shortly.

In comparison with the futile search for the luminiferous æther, the speed of light was quickly established, with astonish-ing precision, to be about 300 million m/s in empty space (what we call the vacuum). Today, the speed of light has been measured to be precisely 299,792,458 m/s.[1] The fact that light *always* travels at the same speed through empty space is a well-publicized piece of scientific trivia, the kind of thing that most adults will have heard and take for granted. But what exactly does it mean? At first glance, it appears that nature has been kind to us for once because this constancy indicates a certain simplicity. However, to fully comprehend the significance and peculiarity of this law, let's imagine an analogous yet more mun-dane scenario in which a car is traveling at the speed of 30 km/h.

Imagine a pedestrian standing on the sidewalk, watching a car drive down the street at 30 km/h, followed by a bike that is traveling at a comfortable speed of 20 km/h. From the point of view of the cyclist, the car is traveling at a speed of 10 km/h faster than her. A priori, it does not seem like it would take too much effort for even a casual biker like myself to accelerate and

---

1. If you wonder how precise this value is, the answer is actually infinitely so. The speed of light in the vacuum is now considered to be such a fundamental quantity that it precedes our definition of distance, and the notion of what a meter is descends from it. A meter is now defined as the distance traveled by light in the vacuum in 1/299,792,458 of a second.

speed up to 30 km/h, allowing the car and the biker to travel side by side. From the point of view of a pedestrian, the car and the biker would then seem to be traveling at the same speed. From the point of view of the biker, the car would be moving neither faster nor slower than her: the car's relative speed to the bike would vanish. At least that's what we would infer based on our instinctive understanding of how speeds should add up, an idea that should be familiar from our encounter with Galilean relativity in the introduction. Indeed, precisely this intuition about adding velocities was implicit in Galileo's discussion of an observer enclosed in the cabin of a moving ship along with butterflies.

Following this logic, the scientific consensus in the second half of the nineteenth century held that the speed of light through the luminiferous æther would only be the same for specific observers at rest relative to the luminiferous æther. Anyone moving through the æther would measure a different speed. Our cyclist, for instance, should measure a slightly different speed of light than someone standing on the sidewalk or sitting comfortably in her living room. Of course, the difference between what the cyclist and her sitting colleague would see is ridiculously small when compared to the actual speed of light— about eight orders of magnitude smaller—and not a difference that could have been measured using nineteenth-century technologies. On the other hand, if we consider the effect of faster speeds, such as the motion of the Earth as it orbits the Sun, we should be able to detect interesting effects.

As you sit here reading this book, you are, together with the whole planet beneath you, whooshing around the Sun at a very respectable speed of 30 km/s in one direction. There isn't much certainty in our lives or about the fate of the Earth anymore, but I still expect that in half a year the Earth will be on the other side of the Sun whooshing in the opposite direction. This means that if Galileo is correct, measuring the speed of light right now *should* yield a result that differs ever so slightly from what we will observe six months from now or what we would have measured six months ago. As the Earth also continuously spins about itself while orbiting the Sun, and the Sun itself cruises through space around the center of our galaxy (at about 230 km/s), our measurement of the speed of light should continuously vary and depend on which direction we are looking at. Or so we would conclude if we followed our Galilean instincts about how speeds combine, with the natural expectation that light propagates at a fixed speed *within* a luminiferous æther like ocean waves on water.

Michelson and Morley's Infamous Failed Experiment

Near the end of the nineteenth century, Albert A. Michelson, a physicist at the Case Institute of Technology in Cleveland, Ohio, and Edward W. Morley, a chemist at the nearby Western Reserve University, set out on a mission to detect the effect of the luminiferous æther wind (this was before both establishments merged to form what is now Case Western Reserve University, the university that my husband and I would join some 135 years later). The experiment they devised required building one of the first interferometers, an ingenious device that is now used in a wide range of scientific fields.

To understand how this experimental setup works, imagine flying back and forth between two airports. Under calm conditions with no winds, and ignoring traffic and Earth rotation, each leg would take the same amount of time. Now what happens under windy conditions? If the wind's direction is transverse

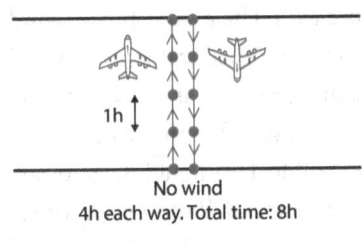

No wind
4h each way. Total time: 8h

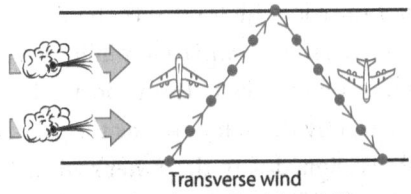

Transverse wind
5h each way. Total time: 10h

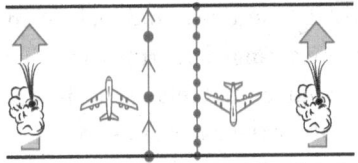

2.5h with tailwind—10h with headwind
Total time: 12.5h

FIGURE 1.1. Delays incurred when flying with very strong winds. Light traveling within a luminiferous æther in motion would follow a similar pattern.

(perpendicular to your path), as shown in figure 1.1, the plane will need to compensate its direction for it, so the time it takes each way will be slightly longer. If, instead, the winds blow in the direction of our initial journey, we gain time with the tailwind and lose some when we encounter a headwind on our return. You might expect that the time we gain from the tailwind simply compensates for the time we lost from the headwind, so our total travel time would be unaffected by wind. Alas, headwinds in fact add more time than tailwinds save. Traveling back and forth in the same direction as the wind involves longer delays than traveling orthogonal to the wind.

Michelson and Morley's light interferometer uses this insight to detect the presence of "light wind," or luminiferous æther. Imagine that at one airport there is a source of light and at the other airport a mirror, which reflects the light back. If light propagates at a fixed speed *within* a luminiferous æther but the æther was in motion relative to the airports, the æther would act on light in exactly the same way as the wind does on our aircraft. If light is aligned with the æther's wind, the roundtrip travel time will be longer than if the æther's wind is perpendicular to the light path. To be a bit more precise, the interferometer splits a beam of light into two, and each beam is then sent down two arms that are arranged in orthogonal directions. At the end of each arm is a mirror that reflects the beam, returning it along the same path where the two beams are eventually recombined. If both beams arrive back at their original point of departure at the same time, they are in phase: the peaks and troughs of their wave profile will match precisely, and both beams will add constructively. But if one beam lags behind the other one—for instance, because the motion of the æther slows one down more than the other—they will no longer perfectly team up when they recombine. They will be out of sync or out of phase, and the amplitude of the resulting signal will be weaker due to destructive interference. Any modulation in the signal is thus a telltale sign that the effective speed of light is changing along one direction of the æther wind (or, equivalently, the motion of the Earth through the æther).

Figure 1.2 shows a sketch of Michelson and Morley's setup. Remarkably, the entire apparatus was attached to a huge block of sandstone, which was floating in a trough of mercury in the basement of Michelson's laboratory. The experiment was thus free to rotate and explore all possible ranges of orientation, including that aligned with the æther's wind.

The results of the Michelson–Morley experiment, published in November 1887, revealed what we now take for granted but seemed impossible at the time: the speed of light in the vacuum is always measured to be *exactly the same* regardless of how fast the actual observer is moving. Michelson and Morley found absolutely no evidence of any change in the speed of light, despite the undeniable fact that the Earth and hence their experiment was moving through space at different speeds and in different directions throughout the process.

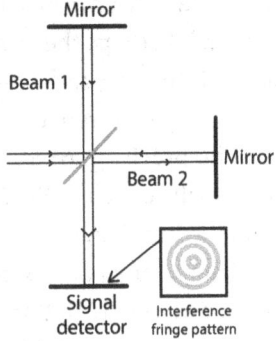

FIGURE 1.2. Sketch of the Michelson–Morley interferometer. If the speed of light is the same along every direction, both beams would combine perfectly synchronized at the screen. If the speed of light in one direction is effectively slower than the other because the Earth is moving through the luminiferous æther, the signals will be out of sync when they recombine.

Only two possible conclusions could be drawn from the results. Either, by some incredible and unexplained miracle, the luminiferous æther remained at rest relative to their laboratory, perhaps implying that Cleveland is the center of the Universe . . . or that the concept of luminiferous æther through which light propagates must be abandoned forever. Given this choice, it is not surprising that the latter possibility eventually gained traction, although at the time the answer was not so clear-cut. The difficulty of this choice was initially compounded by the fact that Michelson's own personal life was fraught with scandal. Perhaps the vast amounts of mercury with which he came into contact during his experiments also might explain the events that were to follow.

The precision of Michelson's optical experiments would eventually earn him the Nobel Prize in Physics in 1907, making him the first American laureate in his field. In light of this later acclaim, many have forgotten the darker side of this brilliant senior scientist. But at the time of the famous Michelson–Morley experiment, Michelson was only thirty-five years old, and he found himself in a bit of a scandal. Michelson was accused of taking "improper liberties" with his housekeeper, who together with her aunt was attempting to blackmail him. Rather than succumb to his blackmailers, Michelson arranged for them to meet him at his laboratory at the Case Institute on the morning of October 10, 1887 (the day before Michelson and Morley were slated to present their findings for the first time at Cleveland's Civil Engineers' Club). With Morley present as a witness and a detective hidden in the laboratory, Michelson asked the blackmailers to repeat their demands. The blackmailers were arrested on the spot—and Michelson himself was arrested two days later. So the announcement of the experiment's results had to wait until December.

In the end, everyone's charges were dropped, but the events nevertheless had an impact on the trajectory of science in the late nineteenth century. Michelson and Morley had planned to repeat their experiment that autumn to ensure that their findings were not simply the result of an anomalous cancellation caused by the specific position of the Earth in the summer when the experiment was first performed. In view of the previous personal events, their follow-up never occurred, at least not as Michelson and Morley had originally intended. Instead, for the next twenty years physicists came up with all sorts of convoluted explanations for why the Michelson–Morley experiment saw no effect, proposing, for example, that the æther was being dragged along by the Earth. Eventually, however, their experiment was repeated with increasing accuracy by multiple people, including Case faculty member Dayton Miller, who would go on to chair the Physics Department.

As unlikely as it first seemed, the scientific community had to finally accept the perplexing reality that, no matter how fast and in which direction we are moving, we always perceive light to be moving at the same speed with respect to us. More generally, whether we are sitting on our sofa, flying at more than 3,500 km/h on the world's fastest jet aircraft, or perhaps even sunbathing on a comet zooming past the Earth at 100,000 km/h, we always see light as traveling relative to us at exactly the same speed. This contradicts the Galilean intuition that as we move along the direction of someone (or some particle, or some wave), we should measure a slightly different relative speed. It was a young Albert Einstein who, in 1905, finally made the conceptual leap and discarded the luminiferous æther entirely. We now understand that rather than propagating through a medium, light is not attached to anything; it needs only space and time to make its way through the Universe.

Today, the Michelson–Morley experiment is widely regarded as the most famous "failed" experiment. Their failure to detect a signal paved the way for some of the past century's greatest breakthroughs. Remarkably, interferometers based on essentially the same design are now routinely used, though the mercury bath has been traded for quasi-monolithic suspensions, the size of the arms has been scaled thousands of times, and a two-man team has been replaced by thousands of scientists from hundreds of institutions. The interferometers currently used by the U.S. Laser Interferometer Gravitational-Wave Observatory (LIGO), the Virgo European Gravitational Wave Observatory near Pisa in Italy, and Japanese Kamioka Gravitational Wave Detector (KAGRA) collaborations now produce results that are approximately eighteen orders of magnitude more precise than that of the original Michelson–Morley experiment! Nevertheless, the underlying concept behind these interferometers is the same. An experimental design that was once conceived to furnish proof for our Galilean instincts would turn out, a century later, to be instrumental in the detection of gravitational waves, "proving" general relativity instead.

Rotations in Space and Time

The realization that the speed of light is universal for all observers forced scientists to reconsider how to describe the fundamental nature of our Universe. Ultimately, it would require that we merge the very notions of *space* and *time* together into a single unifying framework, one that transcends simple expression in any human language. The new mathematical language developed and used to describe the world would involve overcoming our natural scientific instincts to rebuild our laws of logic. Though it was Albert Einstein's incredible 1905 proposal of what we now call the theory of special relativity that is most immediately responsible for this conceptual revolution, this shift in thought would not have been possible without the brilliant insights of many others in the international scientific community, particularly the Dutch physicist Hendrik Lorentz, the French mathematician Henri Poincaré, and the German mathematician Hermann Minkowski (who also happened to be Einstein's teacher in Zurich).

When we compare what a pedestrian and a cyclist perceive, we accept that the cyclist is moving with respect to the pedestrian and thus their respective positions in space evolve differently. However, when we apply Galileo's rule about how to combine the speeds, we make one understandable but critical error: we assume that everyone shares the same perception of time. If the cyclist goes on a one-hour ride, we assume that both the cyclist and the pedestrian would agree on what the notion of "an hour" means; we assume that, even though their relative positions in space evolve differently, they both see and feel the passage of time in the same manner. In doing so, we implicitly treat space and time differently. Time, according to Galileo and Newton, is an absolute concept whose passage is the same for all observers regardless of their motion. Yet this idea is incompatible with the universality of the speed of light. And because all experiments agree that the speed of light is the same for all observers, we must abandon the idea that the flow of time is universal.

Einstein's breakthrough in the formulation of special relativity was to realize that time, like space, is affected by motion. The driver of a fast-moving car's perception of time differs from that of a slower-moving cyclist, which differs still from that of a pedestrian. At everyday speeds, these differences are imperceptible, but as we approach the speed of light they can become large. Crucially, the difference in our perception of time as we move at different speeds in various directions affects how speeds add up such that when we measure effects that occur at the speed of light, we all come up with the same speed, irrespective of our own relative motion. If this seems abstract, it is because our natural intuition has broken down by this stage.

As humans, we have evolved to be particularly aware of what happens at relatively low speeds, where our Galilean intuition is correct. After all, none of our predators (at least none that we know of so far) hunt at speeds close to that of light. Nevertheless, the mathematical proof behind this idea only requires us to be a little more flexible in our perspectives. Indeed, once we accept that space and time are not two disconnected concepts but should be wedded into a single one, which we call spacetime, what happens as we approach the speed of light becomes much clearer. As Hermann Minkowski brilliantly realized, comparing the perspectives of the pedestrian and cyclist is just like performing a rotation—not a rotation in *space*, but a rotation in *spacetime*.[2]

Imagine you start your life somewhere in the Southern Hemisphere, and you learn to identify the various stars and the constellations they make in the sky. You know that how they move through the night sky depends on the seasons, knowledge that allows you to navigate in the dark and make sense of what surrounds you. At some point, you relocate to the Northern Hemisphere, where you quickly learn how to enjoy a new culture, a new language, new friends, new dishes, and new traditions. You can embrace all these changes with the confidence that some elements of reality will remain constant: a day is still twenty-four hours long, and rainbows still shine with the same colors in the same order. Surely, the stars in the sky cannot have been affected by your decision to take a job somewhere up north. And yet, by changing your position on the surface of the Earth, the sky above you will have rotated and will look quite different. Initially, the

---

2. To be more precise, those are rotations in *Minkowski spacetime*, a flat, four-dimensional geometry made up of the three dimensions of space and the dimension of time viewed geometrically. These spacetime rotations, which mix space and time, are nothing other than what Poincaré would call *Lorentz transformations* and are a special example of the full set of *Poincaré symmetries* enjoyed by Minkowski spacetime.

stars' respective orientations would seem twisted, though soon enough you would be able to identify familiar constellations and make sense of the sky for what it has always been.

The situation is similar when we compare different people moving at different speeds. In that case, it isn't merely that observers move relative to one another in space, but that what they identify as the directions of space and the time are rotated relative to one another. Two people can have different perspectives, but neither possesses the "correct" picture of the sky. Their views may be rotated in comparison to each other's, but they are equivalent. From the point of view of physics, everyone is equivalent, which is why we all observe the exact same laws of physics and measure the exact same speed of light, irrespective of how we are oriented or how fast we move with respect to one another.

This universality means that no matter how fast we try to accelerate, light will always be traveling at the same incredibly fast speed with respect to us. There is no catching up with it, and the speed of light hence sets a maximum speed limit that can never be overcome. This idea, that light sets a universal speed limit, is often introduced as an assumption underpinning the laws of relativity, and, historically, this is how Einstein derived special relativity. However, the opposite is closer to the truth: it is the unavoidable unification of space and time, combined into the new geometric structure of spacetime, that informs us that nothing can ever travel faster than light.

The Equivalence Principle and Universality of Gravity

In the development of special relativity, Einstein realized that the universality of the speed of light is not just an innocent coincidence. It is nature's way of communicating its elegant fundamental symmetry: when it comes to describing the laws of physics, all observers are equivalent irrespective of how we rotate them and push them apart. Once we realize that, we may begin to wonder if the *universality of gravity*, highlighted by Galileo and encoded in Newton's law of universal gravitation, might itself be another hint from nature.

Legend has it that Galileo proved the universality of gravity by dropping two objects of different masses from the iconic leaning tower of Pisa and witnessing them fall to the ground at the same rate. Whether or not this legendary free fall experiment really took place is still up for debate. However, more recently, by dropping a feather and a hammer on the Moon, Apollo 15 commander David Scott unequivocally demonstrated that objects fall at the same rate regardless of their shape, mass, density, taste, or color, at least when friction is negligible (as it is on the Moon, thanks to the lack of atmosphere). The force of gravity pulling us all toward the surface of the Earth, or feathers and hammers toward the surface of the Moon, acts on everyone and everything in the same *universal* way. Could this be yet another innocent coincidence? Or, as Sherlock Holmes might indicate, "the Universe is rarely so lazy" after all.

To appreciate the significance of universality as captured by Galileo, we can compare the laws of gravity with the laws of electromagnetism. Rub a party balloon on your head. As you gently remove it, your hair will be attracted to the balloon. This phenomenon occurs because, during the rubbing process, you transferred some of the electrons from your hair to the balloon, and when you pull the balloon away, your positively charged hair is attracted to the negatively charged balloon. Simple electrostatic child's play! Now, if you tried the same trick with a piece of cheese instead of a balloon, chances are that, apart from making your hair smell like cheese, not much would happen;

because a piece of cheese does not easily transfer its electrons, it does not become particularly charged. Crucially, the acceleration resulting from the electric force depends on the charge of the object: whether an object is electrically neutral, positively charged, or negatively charged, and by how much, makes all the difference. Gravity, by contrast, is radically different—it is completely unbiased! We all experience it in the same way, whether we are a person or a planet, a black hole or a balloon, a hammer, a feather, a piece of cheese, or a pumpkin seed.

In thinking through this analogy, you may be tempted to interpret the mass of particles as their "charge" with respect to gravity: the more massive a particle, the stronger the attraction. Behind this thought is an understanding of "mass" as an innocent concept that can be measured by a simple bathroom scale. When you stand on your bathroom scale in the morning, you inflict your own weight on the scale, which is nothing other than the gravitational force exerted on you by the Earth: the greater your gravitational mass, the stronger the gravitational force exerted on you and measured by the pressure you inflict on the scale.

To see why this doesn't provide the full picture, consider the following thought experiment. Imagine that, together with a hammer and a feather, Commander Scott had brought his bathroom scale with him on the Apollo 15 mission and weighed himself on the Moon. The *weight* displayed on the scale on the Moon would be quite different than that displayed on Earth. If the scale read 100 kg on the surface of the Earth before taking off, it would indicate about 16 kg four days later upon reaching the surface of the Moon. At this point, you may well be wondering why anyone would want to become an astronaut if the food onboard the spacecraft was truly so terrible that you would lose 84 kg in four days. But if flying to the Moon actually resulted in such a drastic change of your body mass, I am sure that dieting companies would have already developed Earth-to-Moon weight loss programs. In fact, your gravitational mass (or "charge" with respect to gravity) is exactly the same whether you are on the Earth, on the Moon, or in the middle of empty space, just like the electric charge of an electron is always the same.

The reason the scale displays different numbers on the Earth and the Moon is not because your gravitational mass has changed. Rather, it is because the gravitational mass of the "object" the scale stands on—the Earth in one case and the Moon in the other—differs by about a factor of 100. Because the Moon's radius is about four times smaller than that of the Earth, Newton's law of universal gravitation tells us that the force of gravity on the surface of the Earth is about $100/4^2$, or ~6.25 times greater than it is on the Moon's surface. This explains why your weight, as measured by the scale, is so much less, as well as why it is easier to jump higher on the Moon than on Earth, despite your gravitational mass remaining unchanged.

Our gravitational mass is what tells us how gravitationally attracted we will be to, say, another planet. But if we want to understand how gravity affects us, we must keep in mind that attraction is not everything; how we respond to it is just as important. And that response is dictated by our *inertial mass*. Waving goodbye to David Scott and coming back to our daily routine on Earth, every morning I walk my three daughters to school. During our journey, I am often entrusted with carrying their schoolbags loaded with notebooks, library books, friendship toys, water bottles, and pebbles collected along the way, giving me a chance to contemplate the significance of inertial mass. The inertial mass of those schoolbags determines how difficult it is for me to drag them (or accelerate them) from home to school. Getting to school faster would require accelerating the inertial mass of those schoolbags on my back to a greater speed and in principle would "simply" require more effort (and sweat) on my part. According to Newton's second law of motion, the force required to accelerate a body is proportional to its inertial mass.[3]

The art of flying is to perfect the interplay of inertial and gravitational masses. With our seat belts fastened and tray tables in their upright and locked position, we experience a small thrill as the plane zooms down the runway, the engines spinning on full blast to accelerate the plane's inertial mass to a sufficient horizontal speed. The engines' job is not to counteract the force of gravity but rather to propel this enormous inertial mass horizontally to the point where the pressure from the flow of air beneath the wings acts vertically to overcome the gravitational force that pulls the aircraft down.

Let us imagine, for a moment, that it was possible to engineer two aircrafts with the same gravitational mass so their weight on Earth would be the same but with one aircraft carrying half the inertial mass of the other. Comparing the two, it would be much easier and twice as cheap to accelerate the inertially lighter aircraft. Given this, when we check in for a flight, what should really matter is not the weight of our luggage, which is related to its gravitational mass, but rather its inertial mass, which encodes how much effort it takes to push and accelerate it. Rather than weighing our luggage at check-in, we should really be running a sprint race with our luggage to determine its inertial mass. Or,

3. As we proceed to uncover the layers of fundamental physics, we will later discuss the significance of the mass of fundamental particles. In doing so, we will be talking about their inertial mass. To put it simply, dragging around those massive particles will require more effort. The greater the inertial mass, the greater the push required to accelerate a massive particle. The lighter the inertial mass, the farther you can travel with it.

at the very least, this is what we would be doing were it not for the *equivalence principle*.

A priori, gravitational and inertial masses would seem to be two independent concepts, just as an object's electric charge is independent of its inertial mass or, for that matter, its color. These are simply distinct physical properties. Yet, to date, every single experiment that has been conducted here on Earth, as well as throughout the solar system, has pointed to the same conclusion: the *inertial* and *gravitational* masses of every single element of matter are identical within a precision of up to 10^{-15}.[4] Even though their meanings differ, the concepts of inertial and gravitational masses can be treated as being completely *equivalent*. Einstein's insight was to recognize that, rather than being a mere coincidence, this equivalence was one of the pillars of gravitation, what we now know as the "equivalence principle."

On a practical level, the equivalence principle implies that we can trade one notion of mass for the other. We can infer our inertial mass by weighing ourselves even though, technically, the scale only probes our gravitational mass. Beyond making the airport check-in process slightly less chaotic, the broader implications are far more profound. For one, the equivalence between inertial and gravitational mass means that these two quantities cancel each other when responding to gravity. As we drop a hammer and a feather on the Moon (or on the Earth, neglecting the friction from the air), the hammer experiences stronger gravitational attraction, but it also has larger inertia. The feather, on the other hand, experiences weaker gravitational attraction, but in turn its inertia is also smaller, and so the feather responds

4. This means that if an object has an inertial mass of exactly 1 kg, its gravitational mass should be between 0.999999999999999 kg and 1.000000000000001 kg. All experiments and observations performed so far suggest that the inertial and gravitational masses are always exactly the same.

more efficiently. The equivalence principle implies that the effects from the gravitational and inertial masses cancel out, and the hammer and the feather fall to the ground at the exact same rate—as does everyone!

Einstein's remarkable achievement was to realize that this equivalence of mass meant that *gravity* (which cares about gravitational mass) should ultimately be related to *motion* (which cares about inertial mass), leading, via one of the greatest conceptual leaps in science, to the end of the Newtonian gravitational force and the birth of a much more fundamental description of gravity—namely, through the curvature of spacetime.

The universal speed of light and the universal nature of gravity, far from being coincidental, are the first signs that, at its most fundamental level, nature is perfectly symmetric and universal. Recognizing these universalities provides us with an opportunity to reveal a new layer of physics, opening a small crack in the structure of knowledge through which we can glimpse a new pathway toward making sense of reality. Peering through this crack will lead us to Einstein's theory of general relativity. And as we delve further into this opening, new challenges will emerge to the very foundations of general relativity itself.

CHAPTER 2

The Gravity of Our Curved Reality

Having moved around constantly throughout childhood and adult life, mixing continents and cultures, one of the first questions I will inevitably be asked upon meeting someone new is where I or my daughters come from. The question always makes me laugh, somewhat embarrassed not to have come prepared with a simple answer, ready to respond to this perfectly simple question. There is never an easy answer, though, so "from all over" is what my American-British-Canadian-Swiss daughters usually settle for, leaving our interlocutor somewhat frustrated at not being able to pin us down to a specific point on the map. Of course, a part of me wishes I could be more specific, that I could trace a straight path from my point of origin to my present circumstances.

When I turned ten, I moved from Switzerland to Madagascar with my parents and four siblings. Aboard the overnight flight from Zurich to Antananarivo, as I left behind almost everyone and everything I had known for the past few years, I was struck by the Moon silently following us on our journey. After landing, along with the excitement of a new life and new adventures to discover, it became clear that my entire existence was nothing more than a patchwork of events in space and time, and it was up to me to bridge between them. I soon came to realize, however, that while these various patches may appear disconnected and independent from the outside, from the inside they are connected by the course that my life has taken, by the story that I chose to create—falling, flying, straight as an arrow through existence's complex manifold.

Throughout this journey, the Moon, and indeed the whole sky, remained my faithful companions. No matter where I was, I knew I could always count on them, just like family. Naturally, over the years the desire to get closer to them, understand them, and in turn be there for them grew stronger. They became a dream, perhaps even a fixed point that I knew I wanted to reach, a desire that would define every one of the steps on my journey. For as long as I can remember, I dreamed of exploring the sky and becoming an astronaut. And rather than being a mere passing fancy, this dream guided my professional decisions for decades to come. Pursuing my passion for the sky would take me all over the globe, from Antananarivo to Lausanne for my master's degree, to the École Polytechnique in Paris for another master's, to NASA's Jet Propulsion Laboratory in California, then back to Switzerland and on to the University of Cambridge for a PhD in theoretical cosmology. From there I would go on to live in Montreal, Waterloo, Geneva, Cleveland, and finally London, with stays all over the world, from Cape Town to Kyoto, from Quito and Lombok to Princeton.

To an outside observer, bouncing between new countries every few years or even every few months may look like an uncertain and meandering career path. Yet this was perhaps the most direct way for me to pursue my desire to explore the Universe. In retrospect, there is no such thing as a twisted life path in a straightforward world. Rather, we all follow a straight path through our lives, constantly adapting to the curved reality of the world in which we find ourselves. As we freely fall through our lives, it is the unique way that we connect the various events together that makes our path so special and precious.

The Thread in the Fabric of Spacetime

While my globetrotting journey had its twists, I never stopped fantasizing about traveling to outer space and experiencing the sensation of weightlessness, of tracing a path as a "free-falling observer." People sometimes think of weightlessness in space as the complete absence of gravity, but this isn't quite right. Even if you are fortunate enough to fly in outer space, there is no escaping gravity's universal attraction. In fact, as long as you remain "close enough" to the solar system or any other astronomically massive object, there is nothing—no amount of money, no conceivable technology—that will ever persuade gravity to interrupt or disrupt its continuous attraction. To free ourselves from the gravitational attraction of the Earth, the Sun, and our local cluster of galaxy, one could simply travel for a few million years (at the very least) to reach the middle of the nearest cosmic void in search of an environment with as little gravitational attraction as possible from any surrounding massive galaxies. However, if this amount of time seems just a tad too long, there is an easier way to enjoy a sensation similar to weightlessness: surrender yourself to gravity, falling freely, in complete harmony with it. Weightlessness and free fall feel akin to each other, and the latter, of course, is the only way that any human has ever experienced the sensation of weightlessness.

Whether we are well-rooted on the ground, skydiving through the air, or orbiting the Earth on the International Space Station (ISS), we never actually lack mass or weight. Instead, it is the ability to simply fall freely without interruption that gives us a feeling of weightlessness. Imagine jumping off a diving board. You will experience near-complete weightlessness for a second or so. As you return for a second jump, this time make sure to savor the moment by closing your eyes and blocking out any external distractions. Can you tell whether you are falling toward the Earth's surface in that split second of free fall? Or are you perhaps on a space mission plummeting to the surface of a newly discovered habitable planet, waiting for your parachute to open? Or are you on a space station orbiting the Earth, where your accelerated motion compensates for the Earth's gravitational pull? Or are you, in fact, all alone in outer space, so very far away from any planet and star in a Universe so old, empty, and flat that you can truly experience the total absence of gravitational attraction? During that split second of free fall, it would be impossible to tell. The sensation of free fall is identical (at least locally) to what we would experience if neither the gravitational pull of the Earth nor any other inertial force were present, hence achieving the perfect illusion of a zero-gravity environment.

This equivalence between accelerating in free fall in a gravitational field and being at rest in the absence of gravity is a direct consequence of the equivalence between the inertial and gravitational mass that we encountered in chapter 1. It can be stated in a different but related way. An observer locked up in the cabin of a ship that is at rest under the influence of gravity will have the same experience as an observer locked in the cabin of a ship that is undergoing constant acceleration in empty space far from any other bodies' gravitational influence.[1] The inertial force caused by acceleration acts like artificial gravity.

---

1. This is true until, tired of bouncing in weightlessness, the observer decides to compare the motion of multiple free-falling coffee beans. As we will see in chapter 3, it is precisely in comparing different points that one can learn something about gravity.

In my dreams, I often return to those wonderfully serene moments when I was swinging on an Amazonian hammock as a child, and I feel for a moment as if I am floating freely out of space, time, and gravity . . . until the persistent pull of gravity brings me back to the ground and to reality. Or does it? If we close our eyes again and focus our attention on the pressure we feel from that which supports us—whether the strings of the hammock, the bottom of our chair, or the floor pushing against the palms of our hands as we attempt a handstand—how can we be certain that the pressure we are experiencing is really originating from the gravitational pull of the Earth? As we close our eyes, can we be certain that we are not actually in empty space and that the sensation we are experiencing is not the result of our space rocket pushing us with constant acceleration? Or that we are not in an outer space facility that is spinning around itself so as to precisely simulate the artificial feeling of Earth's gravitational attraction? In short, the answer is no. From our own personal experience, there will be no difference between these wildly different scenarios.

As we observed in chapter 1, the key difference between gravity and other forces, such as electromagnetism, is that gravity affects everything and everyone in the exact same way. Whether you are a single atom of oxygen in the ISS, an astronaut, or even light itself, the effect of gravity on you is the same—and locally it can be mimicked for everyone and everything in the same way by shifting to an accelerating frame of reference. But if we can so easily remove or reproduce its effects by simply "changing perspectives" to an accelerating frame of reference, how can we think of gravity as a real force?

For some, the concept of gravity is associated with the image of an apple falling from a tree toward the surface of the Earth before hitting Sir Isaac Newton's head. For others, the picture of an astronaut in orbit may come to mind. In either case, whether you are an apple falling freely toward the surface of the Earth or an astronaut in orbit, you will be falling. As you fall, you will navigate Earth's gravitational field. Yet at every instant during the fall, you will experience the same feeling of zero gravity. This means that at every point, at every moment, we could always describe what happens locally as if there were no gravity. So how and where does gravity ultimately come into play? The answer is simple: gravity comes into play in how these different points in space and time connect with one another. As we have seen, locally gravity is meaningless, so the only way gravity can manifest itself is through the connection between local regions. Gravity is the thread that connects the patchwork of regions of space and time together. It is in this sense that gravity can be thought as being related to the very fabric or structure of spacetime.

If various patches in the fabric of spacetime are connected in a direct and straightforward way, then the "manifold" we live in will be flat.[2] But as soon as there is anything present on that fabric—be it the tiniest speck of dust, the smallest particle, massive or not, any form of energy, even pressure—this fabric will bend. Just as pressing on a patchwork blanket affects the tension of the stitches that hold together the various pieces of fabric, the presence of anything on a spacetime continuum will affect that manifold by altering the connections between the various points, curving the very structure of space and time in the process. It is through this curvature that any object (material or not), creates a gravitational field around itself and thus communicates its presence to the rest of the Universe.

2. A manifold is a smooth space(time) geometry. It can be a surface, like the surface of the Earth, but it can also be a three-dimensional space, such as the present Universe in which we live right now, or even a four-dimensional spacetime.

The fabric of spacetime is often compared to the surface of an elastic trampoline. If you leave a massive ball in the middle of the jumping mat, it will sink down and curve the surface such that other objects on the mat will migrate toward it. Although this analogy provides a useful mental image with which to work, it is flawed for several reasons. For one thing, we really need to imagine a curved *spacetime geometry*, not just a curved space. Many of the most interesting insights in physics have emerged from the delicate way that space connects with time (as we have seen when accounting for the speed of light and as we will see below), so if we only contemplate space and not spacetime, we are missing the bulk of what gravity is. Just as water striders confined to live on the surface of the water will never be able to appreciate how fish can swim through the depths of the water, any scientific investigation that confines itself to curved space will remain ignorant of so much that occurs in spacetime geometry.

Slightly more problematic is the fact that the trampoline analogy represents the curvature as embedded within a different dimension. The surface of a trampoline is two-dimensional, and the analogy makes use of yet another dimension—the vertical direction or height of the trampoline—to depict its curvature. But when it comes to the curvature of spacetime, we should imagine it curving within itself rather than curving within another space.

Finally, in the case of the ball curving the surface of a trampoline, other objects move toward the ball at the center not because they are attracted to the ball per se (they are to a tiny degree, but this is utterly negligible), but because they are gravitationally attracted to the Earth and moving closer to the ball allows them to move closer to the Earth. These subtleties simply show that humans haven't—yet—evolved to perceive nature through four-dimensional spacetime eyes and ears. Fortunately, our imagination and creativity, combined with a bit of mathematical skill, can, help us "see" beyond our bodily senses.

A Spacetime Geometry

In fact, our society has become so accustomed to the implications of the curvature of the spacetime in which we live that we could no longer function if we neglected them. We would quite literally lose our way. Though most of us are destined to spend our life bound to the surface of the Earth (perhaps occasionally traveling as high as thirty thousand feet in the air), our everyday life is constantly supported by the help of satellites in orbit around the Earth. And the successful operation of these satellites depends on accounting for the subtle differences in the way we experience the flow of time at these different points, which itself depends on the local curvature of spacetime.

Return to the scene of Apollo 15 commander David Scott dropping his feather and hammer on the surface of the Moon. If Scott were to stay for a century on the Moon—perhaps he found some prime real estate and did not want to risk giving it up—what would feel like a century for him would feel slightly shorter for those of us on Earth—about 2 s shorter, to be exact. In other words, clocks appear to run ever so slightly faster on the Moon than they do on the surface of the Earth. This is "simply" because the spacetime curvature we experience on the surface of the Earth is stronger than that on the surface of the Moon, and this spacetime curvature affects our perception of time. For the same reason, one century on Jupiter would feel like a century plus an extra minute on Earth. We grow older slightly quicker on Earth than we would on the surface of Jupiter.

Within the solar system, this difference in time flow seems utterly negligible at first sight, and the thrill of these effects typically only arises in science fiction–like scenarios when we are sent to spend a minute in the vicinity of a black hole only to discover that, back on Earth, hundreds of years have elapsed. Yet, far from mere science fiction, these effects are omnipresent in our everyday lives. Indeed, as we spend one day on the surface of the Earth, the Global Positioning System (GPS) satellites in orbit some 20,000 km or so above our heads feel one day and a tiny, tiny bit more—38 microseconds (μs) or 0.000038 s more, to be exact.[3] Granted, this is not a length of time that humans have sufficient sensitivity to feel. But light does! In 38 μs, light propagates over 11 km. If we didn't account for the differential effects of the curvature of spacetime between satellites and the Earth, the locations indicated by GPS devices would be off by a dozen or so kilometers every day.

Without a proper understanding of the spacetime geometry in which we live, we would be back to a much quieter world, using paper maps. Holding virtual meetings between various corners of the planet would still belong to the realm of science fiction. As we continue on our straight path through the curved manifold of life, remember that the curvature of spacetime is paramount to even the simplest journey we take.

A Straight Line in a Curved Spacetime

The research work of a theoretical physicist is often portrayed as a lone adventure: sitting at our desk in slippers, hair slightly unkempt, flipping through papers and occasionally standing up

3. There are two separate effects here, one from special relativity and the other one from general relativity. The motion of the satellite slows the GPS clock by 7 μs, but the stronger gravity we feel on the surface of the Earth slows ours down by 45 μs relative to that of the satellite, leading to a $45 - 7 = 38$ μs difference.

to draw incomprehensible symbols on our blackboards while muttering equally incomprehensible mathematical incantations. It is difficult for me to tell whether this is an image that we, as physicists, have somehow unwittingly or subconsciously promoted in order to scare away any potential "perturbations" that might disturb the equilibrium to which we have become accustomed or whether this image is the product of a genuine misunderstanding of what physicists do. In either case, the portrait couldn't be farther from reality. Communication, brainstorming, and the exchange of ideas are critical to creative work and thus to scientific success. As a physicist, I travel the world to share my research ideas with colleagues, who consider, challenge, and even occasionally ridicule my proposals. Most of the time, this process helps me revise, refine, and reconsider every single one of my hypotheses—before ultimately discarding the idea and starting again with better hindsight. True advances in science are almost always hard-won, and the rough-and-tumble process is not for the faint of heart.

Nor is it for anyone who does not like traveling. By the time my first daughter was born, traveling to various locations around the globe to communicate with colleagues had become a normal part of my life and quickly became part of hers. Before her first birthday, she had sat through fifty-one flights, joining me on various scientific exchanges; from her perspective, exploring the curved surface of the world must have seemed as natural as breathing. While I am not proud of the amount of carbon these trips released into the atmosphere—and I would certainly make different decisions today—I will always cherish the global perspective that these trips allowed me, and then her, to develop, and I hope these aspects will continue to be preserved.

These trips also offered insights into how we navigate within the curved structure of reality. At any given point on the surface of the Earth, we may have the impression that the Earth is flat. We know that this is an illusion—zoom out far enough and it becomes obvious—but wherever we are, we can always adapt our perspective so that, locally, space is flat. Curvature only comes to life when we describe how different points on a surface are intertwined and patched together. Only as we travel on the globe's surface—tracing a path between different locations on the Earth—does the curvature become meaningful and reveal itself.

The idea that the Earth is round, rather than flat, first emerged more than two thousand years ago when the ancient Greek mathematician Eratosthenes, who was staying in the Egyptian city of Alexandria, compared its shadows with those of Syene, a city located about 784 km south, precisely on the Tropic of Cancer. For an instant, at midday on the summer solstice, citizens of Syene can experience the Sun hanging directly above their heads. At that precise moment, light rays shine straight through to the bottom of wells in Syene, and its residents could notice light shimmering on the water surface down below. Or so Eratosthenes was told. But nothing of the sort would happen in Alexandria. Instead, Eratosthenes observed that at the same time in Alexandria, light rays from the Sun would strike the ground at a seven-degree angle. From that observation, he inferred an excellent estimate for the curvature of the Earth, illustrating how curvature manifests itself when communicating between different points.[4]

---

4. Syene is located 784 km south of Alexandria, on nearly the same longitude. If the Earth were flat, light rays would strike both cities at the same angle, and shadows of the same objects would appear identical in both locations. Eratosthenes deduced that the Earth had to be curved after noticing a 7° difference in how light rays reached both cities. The radius R of the Earth must be about $R = 784 \text{ km}/\tan 7°$, corresponding to approximately 6,385 km. With the hindsight of two thousand years of wisdom,

The curvature of the Earth affects not merely whether one is able to see into the bottom of a well, but more importantly how we traverse its surface. Imagine, for instance, a trip going straight from the city of Lima in Peru, where I once lived, to the very north of Madagascar (figure 2.1). Lima and the northern tip of Madagascar are located at a nearly identical latitude (about 11.5° south), while the tip of Madagascar is about 13,000 km due east of Lima. If we were to begin our journey in Lima and board a flight bound for Madagascar, we might naively assume that the pilot would simply set the plane on a straight eastbound trajectory. Yet doing so would show little appreciation for how our world fits together. A straight eastbound trajectory on the curved surface of the Earth is not the most direct path, nor is it the shortest journey that would connect the two locations. Such a path, in addition to being longer than necessary, would require the pilot to constantly redirect the aircraft in order to maintain a pure eastbound trajectory. Instead, every respectable pilot would first set a southeast-bound trajectory for about half the way, and then, without ever changing course, the plane would naturally follow a northeast bearing for the remaining half of the journey (assuming little wind disruption and other related effects). Such a trajectory would appear as an arc on a projected map of the Earth, but it is actually the "straightest" line connecting both places.

A straight line connecting two points on a curved space is simply the segment with the shortest distance, which is also the trajectory we would naturally be inclined to follow unless

---

this is remarkably accurate, as we now know the radius of the Earth to be 6,371 km. This difference in angle is precisely reflected in how we now parameterize locations on Earth, with Alexandria's latitude being 31° north and Syene's being 24° north, precisely 7° south of Alexandria.

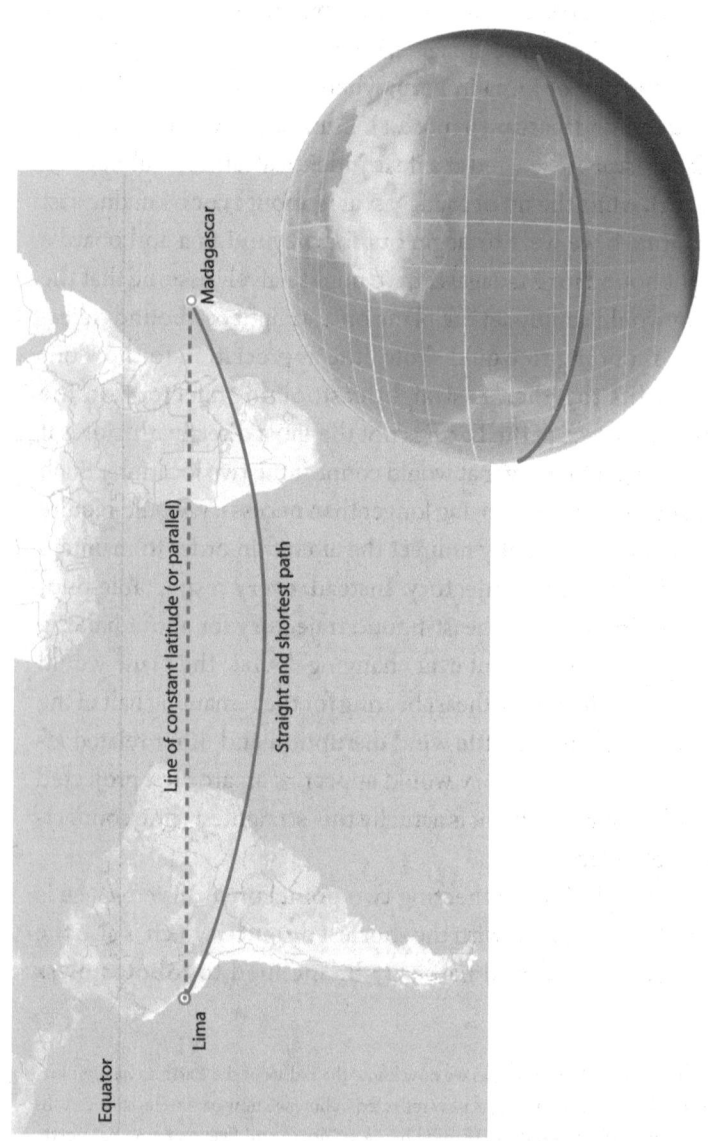

FIGURE 2.1. Straight line between Lima and Madagascar.

something forces us off course. Indeed, if we connect two points by tracing the shortest path between them, then once we have set off on that bearing, we are bound to get to our destination. Any deviation will simply lengthen our journey. Much like my own journey—from Peru to Madagascar, from dreams of becoming an astronaut to studying physics—a straight path through our curved reality will always be the most direct route, even if often it doesn't immediately appear that way.

Our instincts often lead us to associate straight lines with a ruler on a flat piece of paper. But, once again, nature proves to have far more imagination than we could ever have envisioned. The elliptic-like trajectories of planets orbiting the Sun look nothing like straight lines in a flat space. And why should they? After all, the solar system is not a flat space, not even a flat spacetime. The very presence of the Sun, and to a lesser extent the presence of the other planets in the solar system as well as the dust and dark matter, all inexorably contribute to the curvature of spacetime. Because of its ginormous mass, the Sun has the greatest influence on the curvature felt by the various planets in the solar system, and it is the effect of this spacetime curvature that produces their elliptic-like orbits—or at least, that's how they appear when projected on a flat piece of paper. In reality, of course, they are as straight as can be when put in context of the curved spacetime in which they exist. From the perspective of the planets, and from our own point of view as Earth's inhabitants, anyone free navigates through the solar system by confidently bounding straight ahead. Never stopping, skidding, or turning, there is only one way forward, one straight line in this curved space since the birth of the solar system and since each of us was born on this planet.

CHAPTER 3

Gravity and the Force Within

My mother used to joke that if I decided to walk down the middle of the road, it would be easier to move the road than to remove me from it. Now that my own four-year-old exhibits a similar level of unreasonable determination, I cannot help but admire her force of will. Though I occasionally wish her inner drive came with a stronger survival instinct—particularly when she attempts her own experiments with gravity by balancing on the stair bannister—I know that it will serve her well. Looking back on my own life, my aspiration to become an astronaut was almost certainly doomed. After all, the European Space Agency (ESA) selected only six astronauts from a pool of more than twenty-two thousand applicants in 1992. However, once that ambition was unleashed, it quickly became strong enough to transform the obstacles lying in my way into a set of joyful opportunities.

There is no course of preparation that guarantees one will be among the select few chosen for a space mission, but I wasn't ready to let that fact discourage me. As a child growing up in Madagascar, I didn't have ready access to the internet, TV programs, libraries, or museums to learn about space, but I was submerged in something much better: an incredibly inspiring culture and one of the most beautiful environments on the planet. I also had an unobstructed sky I could contemplate every night, incredible biodiversity to challenge my perspectives, and the Indian Ocean in which I could experiment with weightlessness. Under the water, surrounded by thousands of multicolored foreign life-forms, I was fascinated by my ability to control gravity through the simple airflow in my lungs. I started scuba diving as soon as an instructor was kind enough to accept me, and I quickly discovered a whole new Universe, as well as the uncanny feeling of not being able to tell which way was up and which was down. Eventually, I learned how to merge with this new environment and to enjoy one of the most serene experiences imaginable.

As much as I enjoyed my time in the depths of the ocean, it was clear that the best training would take place above the clouds, rather than beneath the waves. I needed to become a pilot. Alas, flying is not a cheap hobby, and it eluded my grasp for many years. However, when I moved to Waterloo, Ontario, as a postdoctoral researcher, I finally had the chance (and could finally afford) to take flying lessons. I would often go early in the morning, braving −20° C Canadian winters, so that I would be back in time for the start of the regular working day. However cold I felt on the runway, that feeling would vanish as the tiny Diamond Katana drew its nose into the air, uncovering a frozen landscape beneath me. As soon as I obtained my single-engine license, I started taking my colleagues for quick flights that stretched from Lake Michigan to Lake Huron, with a glimpse over Niagara Falls at the corner with Lake Erie. On our way to the airport, we frequently engaged in heated physics-related disagreements, and it still amuses me to recall how these would miraculously fade away once we were in the air, opening a new connection in our professional relationship.

After descending from the clouds, I filled my days as a full-time theoretical physicist, focusing on gravity and cosmology. Despite spending most of my time in front of blackboards or in seminars, the time spent in front of my computer was sufficient to see my eyesight decline, and I needed glasses to continue flying. While my myopia was still within the norms for astronaut selection, I didn't want to take any chances if I could help it. It turns out that the eye, like any other muscle in the body, can be trained through proper exercise. I must admit that I was skeptical at first, but over the next few years I put aside some time every day to train my eyes, forcing them to contract and relax. By the time of the astronaut selection eye test, I had successfully corrected my vision in one eye from 20/80 to 20/20, passing with flying colors the part of the examination that initially had worried me the most.

By that time two decades had passed since, as a ten-year-old, I had set my mind on becoming an astronaut. Rumors had begun circulating in 2007 that the ESA would soon launch an astronaut selection round, the first in fifteen years, so it was time to step up my game. I gathered from past selection rounds that the psychological and IQ tests, comprising the most challenging parts of the selection process, were inspired by the tests taken by Lufthansa pilots. Although I didn't have direct access to those tests, I had a sense of the skills they might involve, so I decided to code some software routines that would help me practice and improve my coordination, spatial representation, and memory. As a cosmologist, I have colleagues who are numerical whizzes and can code any program in a matter of minutes—regrettably, I am not one of them. It took me some time, but in the end I successfully created a set of handmade practice tests that I could run on my computer and adapt at will.

I also knew that fitness would be an important factor in the selection process. I never was and never will be a strong runner, so my decision to join the Waterloo running team was perhaps the clearest indication of the strength of my desire to become an astronaut. My partner Andrew, whose support was present in everything I did, even joined me for an early morning run organized by the team on New Year's Day in 2008, despite the frosty Canadian winter. From fitness training to flying, from eyesight workouts to my Lufthansa-inspired homemade training regimen, I was fueled by an inner force. When we put our minds to it, we all have a force within us—the only question is how deep you need to dig to find it.

Gravity, it turns out, is quite similar in this regard. We think of gravity as an unavoidable phenomenon, something into which we literally fall. But what if there were more to it? So far I have described gravity as being unlike electromagnetism and the other forces of nature, corresponding solely to the manifestation of the curved geometry of the spacetime in which we live. And indeed, describing gravity as the curvature of spacetime is useful when considering how the Sun bends light from distant stars, or how the Earth generates the gravitational attraction that keeps us firmly grounded on its surface. However, this spacetime structure is nothing more than nature's canvas, which allows gravity to exist. As we delve deeper into the layers of gravity, we will see that gravity does carry a force within itself—a force that emerges when we contemplate the evolution of bodies held in its sway.

A Feeling of Gravity

What does gravity feel like? If this question seems meaningless, or simply ridiculous, we could perhaps start with something else. If we replace *gravity* with *light*, the answer would certainly be simpler. Almost every life-form on the planet, from primitive bacteria to plants and animals, has evolved to be well-accustomed to feeling or seeing light. In plant leaves, chlorophyll can directly absorb sunlight and convert it into energy. In humans, the retinas in the backs of our eyes are constantly detecting light and converting it into meaningful electric signals for our neurons to process, and the cells of our skin can feel the warmth of infrared electromagnetic waves as they impact our bodies. (This latter fact led to the 2021 Nobel Prize for Physiology or Medicine, which was awarded in part to David Julius for his discovery of receptors for temperature.) Studies show that even primitive beings such as bacteria distinguish between light and dark.

Now, when it comes to gravity can we expect a Nobel Prize in the near future for the discovery of gravity receptors hidden within our skin? Could the answer lie in the vestibular system in our inner ears? This system, associated with our sense of balance, allows us to determine our position in space. Unfortunately, no matter how well-tuned your vestibular system may be, it will not be sensitive enough to probe the unified structure of spacetime and its curvature, let alone to feel the hidden force buried within it.

Perhaps, then, the feeling of gravity will be connected to the receptors for our sense of touch (the discovery of which earned Ardem Patapoutian the other part of the 2021 Nobel Prize for Physiology or Medicine). After all, the pressure inflicted on our body as we fall and hit the ground seems intrinsically linked to gravity. From a young age, we experience bruises and scratches when we fall down, and we come to associate the feeling of gravity with the shock we experience on impact. Gravity is indeed responsible for accelerating us toward the ground, but aside from the dreadful anticipation of what will happen next, there isn't much to feel during the fall itself—just free fall, that feeling of weightlessness (at least in empty space, where there is no air resistance).

What about the unpleasant sensation we feel on impact? Is that gravity? No, not really. That feeling has more to do with the solid nature of the ground and the physicist Wolfgang Pauli's quantum exclusion principle. Pauli's exclusion principle tells us that two electrons (or two protons inside the nuclei of our atoms) cannot be in the very same place at the very same time— or, more precisely, they cannot be in the same quantum state. This implies that we cannot continue along our trajectory indefinitely when another solid matter is in our way. Because the electrons and protons of the ground are already occupying that specific state, it is impossible for our body's electrons and protons to pass through them. This very inability to share space is ingrained within every substructure of every atom of every cell of our body. It is the exclusion principle, not gravity, that is responsible for the stability of matter and the painful break in our fall.

So how can we truly feel gravity? If there is one thing that we have learned about gravity thus far, it is that unraveling its mysteries is not an endeavor that can be accomplished alone. Although Galileo, Newton, and later Einstein did much of the heavy lifting, even they were never working completely alone. Similarly, there is no unique cell in our body, no taste bud nor nerve ending, that would ever be capable of truly sensing and appreciating what gravity feels like by itself. The feeling of gravity requires comparing what occurs at different points in space and time—just as the curvature of the Earth or the curvature of spacetime is only meaningful when comparing what happens at different locations.

Imagine you and your neighbor decide to embark on an adventure together, and both of you set off in the same direction, exactly parallel, and then continue "straight" ahead. If we were in flat space, one without any curvature, the distance between the two of you would remain the same no matter how far you traveled. But the surface of the Earth is not a flat geometry. As you stand a few meters apart and begin your journey in the same direction, each moving straight ahead, the distance between the two of you will vary due to the curvature of the Earth and, in turn, of the space that lies between you.

Assume, for the sake of illustration, that you begin your journey from the equator and proceed northward. The distance between the two of you will gradually shrink as you head on a straight line north (see figure 3.1). Upon reaching the true geographic North Pole (not the magnetic North Pole), you will literally be on top of each other. Of course, for reasons related to Pauli's exclusion principle, one of you will hopefully be reasonable and make room for the other. Our concern here, however, is that despite each of you following a direct trajectory north, the physical distance between you changed as you moved across the curved surface of the Earth.[1] Because your trajectory was affected by the curvature of your environment, your paths were straight and yet curved, parallel and yet they met.

Of course, the surface of the Earth is only a curved space—not a space*time* per se—so our metaphor is not an exact match, and our intuition for how a curved spacetime can affect us remains somewhat limited. Nonetheless, this example already illustrates how the curvature of space(time) affects differently neighboring

---

1. Unless you are located exactly on the equator, the same will be true if you and your neighbor start by moving exactly east (or west). As we saw in the example of flying from Peru to Madagascar in chapter 2, moving due east would require the pilot to turn constantly. Such is life on a curved planet.

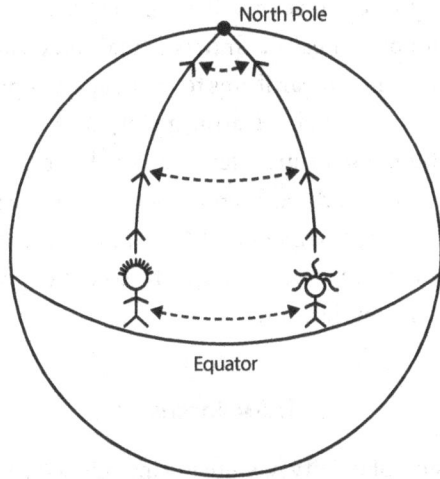

FIGURE 3.1. Following straight lines on the surface of the Earth, one of many examples in which going straight is not straightforward.

trajectories, and it is precisely in this deviation that the force of gravity manifests itself.

Imagine, for instance, that in our prior example we weren't tracing two travelers as they headed north but the cells in our bodies. What if, as we evolved through the curvature of space-time, the paths followed by our left and right hands could diverge significantly? No matter how flexible and eager you are to discover new yoga poses, I am sure there would come a point when this stretching would become a little too much for your taste. If gravity were stronger, we would experience its force in the way the various cells of our body stretch in some directions and squeeze along others: not by falling per se, nor by bruising ourselves upon impact, but rather by sensing the *tidal forces* that pull or squeeze our bodies in different directions as the cells that comprise us do their best to follow the unique spacetime path indicated to them by gravity.

If anything, then, it seems like our best chance at sensing gravity would be to hear it. Gravity's tidal forces would cause the various cells in our eardrums to move apart, long before we would feel, see, or taste it. Hearing gravity on Earth would still require our bodies to be just a few hundred billion billion times more sensitive than presently constructed, so it will most likely be a while until anyone grabs that Nobel Prize. However, if you ever find yourself near a black hole, the sound of gravity could quickly become unpleasant.

Tidal Forces

Our discussion of gravity has already introduced certain topics in greater depth than I do in some of my graduate lectures. Yet we have not, at least directly, discussed Einstein's famous equations. Lest I give the mistaken impression that being a theoretical physicist is all about concepts and ideas, with little concern for calculations, predictions, and numbers, let me briefly stress the importance of Einstein's equations. These equations—or rather what I would prefer to describe as the Einstein-Hilbert action for gravity, from which Einstein's equations can be derived—are absolutely essential to most of the gravitational, astrophysical, and cosmological breakthroughs that have taken place in the past century, enabling space exploration and global communication. It is one of the best-tested theories of nature, in impeccable agreement with all observations and experiments that scientists have undertaken.

Anyone wanting to challenge gravity better be well-versed in the profound insights provided by this Einstein-Hilbert action. The only reason we haven't directly discussed those equations so far is that they are among the most challenging equations to solve mathematically. In fact, Einstein's equations are not just difficult to solve—they are, in fact, impossible to solve *exactly* in all generality! No computer, no matter how powerful and no matter how many times the age of the Universe we let it run, could ever solve them without making some approximations or simplifications. More often than not, our job as physicists is to determine which simplifications or approximations are appropriate in order to make our problems tractable while ensuring that the results remain trustworthy.

One such approximation is the idea that the overall curvature of the spacetime we live in is dominated by the effects of a single body. For instance, in the solar system, we treat the Sun as the body that exerts the main gravitational effect. Every other planet and object in the solar system contributes to a lesser extent, so we consider those to be minor (perturbative) corrections. Within the solar system, such an approximation allows us to make significant progress on a variety of problems and to derive results that agree with observations. But what would happen if we wanted to investigate a more complicated situation, one in which there isn't just a single body that dominates, but where two or more comparable bodies dance gracefully around each other? It isn't long before the mathematical complexities of Einstein's equations come back to haunt us. Already at the level of what may seem like a simple two-body system— for instance, two stars of comparable mass orbiting each other—the time-dependent curvature derived from Einstein's equations becomes too difficult to determine exactly. So we need to make further simplifications (or we have to rely on numerical methods to make the problem more tractable). In physics, even a system as simple as the two-body problem requires considerable effort and skill to solve.

Ironically, another sort of two-body problem has emerged in academia that is almost as challenging to solve, yet it provides us with an interesting analogy for how it feels to be pulled apart by some mysterious force. The two-body problem is a term used in academia to describe a long-standing issue that many couples face. Many academics have a partner who is also an academic, often in a similar field. This is particularly common within theoretical physics: our research often involves focusing on intense, technical topics for years, so it is no surprise that many of us end up developing a deep, special bond with the few other rare souls on the planet with whom we can share our thoughts and ideas.

Of course, if you ask me, I would say that what drew me to my partner Andrew was ineffable and unique, but like countless other couples in academia the two of us met while working on our doctorates, in our case at the University of Cambridge. We were working in neighboring offices at the time, studying the effects of additional dimensions of space on the birth and potential origin of our Universe. Andrew and I often discussed cosmology and gravity together, but it wasn't until we were both sent to a conference in Irvine, California—and spent eighteen hours squeezed next to each other in various planes—that discussions beyond physics triggered the beginning of our lifetime together.

I was just starting my PhD at that time, and Andrew was nearing the end of his. We hadn't been together for three months before he moved "just across the pond" to Princeton, New Jersey, for a postdoctoral position. It was obvious from the start that a long-distance relationship from opposite sides of the globe would be our reality for the decade to come. This is a reality one can deal with for a few years if there is an end in sight, but in academia there is no guarantee of a job at the end of the journey, let alone two jobs in the same country.

After a few years of back-and-forth across the Atlantic, the red-eye London–New York flight was replaced by an overnight bus from New York to Montreal. Then, thanks to some incredible colleagues, we were offered postdoctoral positions together at the Perimeter Institute for Theoretical Physics and at McMaster University, in Ontario, Canada. Our real life, living together in the same location of space and time, lasted three marvelous years before the intercontinental race resumed, first between Waterloo (Canada) and Geneva (Switzerland), and then between Cleveland (United States) and Geneva. Throughout our constant motion, Amsterdam airport came to feel like a second home, and we fell into the habit of picking up small wooden tulips for one another at the airport. A decade later, by the time we had finally both secured a job in the same place and felt we were settled enough to get married, we had collected enough of these wooden tulips to fully decorate our wedding venue. Each of our guests left with one as a small souvenir, planting them in their own homes around the world.

This two-body system, which will hopefully one day find its own natural and elegant solution,[2] provides a perfect analogy for what the real force of gravity feels like. The force of gravity is felt through the manifestation of tidal effects, which is seen in the motion of multiple bodies relative to each other.[3] These bodies

2. I cannot thank enough those incredible colleagues at McMaster University, the Perimeter Institute for Theoretical Physics, Case Western Reserve University, and Imperial College London (who will recognize themselves) for recognizing that behind every two-body problem lies a not-so-hidden two-body opportunity. But my thoughts also turn to all the other two-body systems that are still in search of their own solution, and to those for whom growing apart or unwillingly leaving academia turned out to be the only way forward. Given the wealth of expertise and skill in the scientific community, one would think that for a system as "simple" as the two-body problem, we would already have found a natural and elegant solution.

3. To be more exact, the analogy should really involve four bodies because gravity manifests itself through a "quadrupole," not a "dipole," but we shall leave these technicalities aside for now and come back to them later.

may be squeezed and compressed in one direction while being pulled apart in another direction, testing the very structure of the connection between each body. If you are lucky, the system may recompress and stretch multiple times as it bounces back and forth through its life. And in physics, just as in academia, few are the lucky ones who escape this process as final stable bound states.

A Messenger for Gravity

Leaving behind the mysteries of love and returning to what I claim to understand, the mysteries of physics, let's briefly review what we've covered so far. We have learned that when we fall toward the surface of the Earth, or orbit around it, we are merely following a straight line (or what we call a geodesic) in the curved spacetime induced by the mass of the Earth, replacing any notion of gravitational Newtonian force in the process. Although harder to visualize, but still fundamentally grounded within our everyday technologies, we have also seen how gravity plays tricks on us by affecting how we perceive space and feel the flow of time. Both of these aspects are related to the ambient curvature of the spacetime in which we live, where there is no genuine notion of force to talk about. Yet, hidden within the curvature itself, the concept of gravitational force still exists in the tidal forces that squeeze and pull apart the various points of spacetime— including everything from the cells of our bodies to the fundamental particles at the core of the atoms that make up these cells.[4]

4. Fortunately for us, when it comes to the atoms in the cells, other forces dominate over gravity (for example, the electromagnetic force that holds electrons around atom nuclei or the strong nuclear force that confines quarks into protons and neutrons at the core of every atom in our bodies). But no matter how strong these additional forces are, if you provoke gravity in its most extreme environments, there will always come a point when the underlying gravitational tidal forces will be strong enough to tear the very structure of our body apart.

The instantaneous nature of Newton's law was one of the early signs that Newtonian gravity could not be the final answer, for all attraction involves communication, and communication takes time. It had to be replaced with something more dynamic and responsive. Now that we've embraced general relativity, are we any closer to understanding how gravity communicates? If so, who is gravity's messenger? This may seem like another strange question. As before, it may be easier to approach it by first asking a simpler one: how does communication occur in electromagnetism?

A clever way to determine whether I am carrying a key, coin, paper clip, or other piece of random metal would simply be to ask me. However, in certain situations a rather different, more universal method of communication has been engineered: the metal detector. Pulses of electric current are sent around coils in the detector, creating a brief magnetic field whose electromagnetic influence is transmitted into the cavity. Once this influence reaches any piece of metal within the cavity, it induces an electric current in the metal, which in turn generates a magnetic field that is communicated back to the detector, indicating the presence of the metal object.

The interaction or exchange of information between the detector and metal object is communicated by an electromagnetic influence that travels at the speed of light. When the machine is first turned on, the detector cannot detect the presence of a metallic object in its cavity until light has had the time to travel from the detector to the object and back again, a small yet nonzero amount of time. Unlike Newtonian gravity, electromagnetic forces are not instantaneous! Instead, the electromagnetic force between any two bodies is communicated through electromagnetic waves. Calling them electromagnetic waves may sound cleverly scientific, but these phenomena are quite familiar to us: most of us simply refer to them as *light*.

For metal detectors, the light being transmitted is not visible, at least not to our eyes, but it is light nonetheless. When we walk through even the most reliable metal detector, there is still the silent messenger of light, traveling at a finite speed between whatever piece of metal we may be carrying and the magnets present within the detector itself. This very same messenger, light, is the one responsible for communicating the electromagnetic force between all charged bodies in the Universe, irrespective of the distance between them. If a charged particle were suddenly created, this information (its birth announcement card, if you like) is communicated to the rest of the Universe at the speed of light. It would take a second for an electron located a light-second away (300,000 km) to become aware of the existence of this new charged particle and be attracted or repulsed by it. The metal detector, operating within the space of a few feet, involves time scales so short that they would be difficult to perceive.

After taking this brief detour through our metal detector, we may contemplate developing a similar device for gravity. Everything and everyone is "charged" under gravity, so every particle of our body would technically be responsive to such devices (though the level of sensitivity required to register such tiny forces is far beyond our current and foreseeable capabilities). When it comes to the messenger for gravity, we could give it a similar name to its light cousin. How about *glight*? For now, the scientific community refers to gravity's messenger as *gravitational waves*, the gravitational equivalent to electromagnetic waves. As light or electromagnetic waves are the messengers for the electromagnetic force, glight or gravitational waves are the messengers carrying the information about any gravitational effects we may encounter, from the curvature of spacetime as it is contoured under the mass of the Earth to the presence of tidal forces.

Light, as the name conveniently implies, is indeed *light*—so light that its inertial mass (the mass that determines how difficult it is to set light in motion) is smaller than anything we have yet measured; so light that, for all we know, it is massless. As it takes no effort to set light in motion, it is also impossible to slow it down or stop its relentless propagation, other than by absorbing it. This is why light always moves at the speed of light. Glight has exactly the same properties, at least according to Einstein's equations of general relativity. Gravitational waves are also considered to be massless and maintain a persistent motion as they propagate from one side of the Universe to the other, propagating through space and time at the very same rate as their cousin, light.

Before we contemplate how we can see or hear glight—in other words, how we can detect gravitational waves—it is worth getting a sense for how those waves affect the structure of space-time as they move along. Here, again, it may be instructive to begin with gravity's more familiar cousin. Our eyes are accustomed to capturing light emitted by the Sun, particularly at visible frequencies, but they make little distinction between its different polarizations—the orientation of the oscillation of the light wave. Only when we put on fancy polarized sunglasses can we distinguish between the various ways in which the light we see may be polarized (see figure 3.2).

To better appreciate the different polarizations of light, imagine drawing a wiggly wave moving from left to right. Depending on whether you draw that wave on a piece of paper that is (horizontally) lying flat on your desk or (vertically) taped to a blackboard, you will actually be drawing two different and independent polarizations of light. Maxwell's theory demands that the wiggles of the wave always occur at right angles to the direction of motion, so there are only two possibilities for how those wiggles can appear, which may be thought of as horizontal and vertical

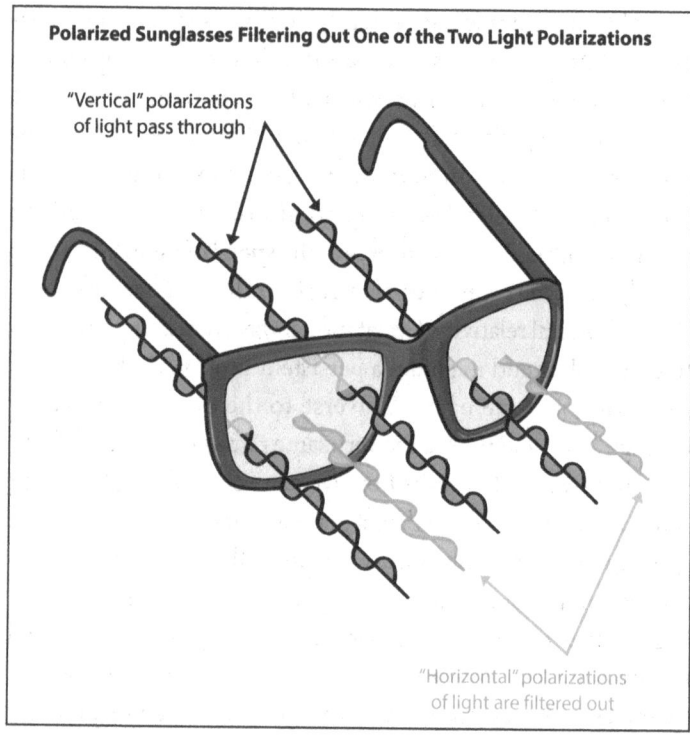

FIGURE 3.2. Different polarizations of light. Polarized sunglasses only let one polarization through. *Source*: Image modified from microscopyu.com.

polarizations (although in more realistic situations light is a mixture of these two, which can be harder to depict).[5] Once set in motion, these two polarizations are free to move independently. They don't need each other to pursue their respective journeys, and if one of them gets absorbed, the other may continue untroubled, which is exactly what happens when you wear polarized glasses.

5. If the speed of light were not fixed (if it could speed up or slow down or even come to a stop), the waves could also occur along the direction of motion, giving rise to a third possibility for how the wiggles of a wave could appear. This would happen if light was not *that* light (if it wasn't entirely massless).

mode ×

mode +

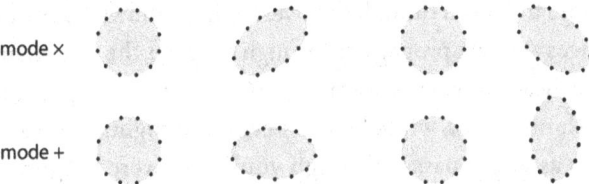

FIGURE 3.3. Polarizations of gravitational waves in general relativity for waves traveling perpendicularly through this page. These waves have been depicted with an extremely large amplitude so that we can "see" their effect with our naked eyes. If we ever detected gravitational waves with such large amplitudes, this would indicate that we are on the verge of a cataclysmic event. We could try running for our lives, but because glight travels at the speed of light we are unlikely to escape. *Source:* Claudia de Rham, "Massive Gravity," *Living Reviews in Relativity* 17, no. 1 (2014): 7.

Glight travels in a similar pattern, but to appreciate the effect of gravitational waves, recall that gravity is intrinsically related to the curvature of the fabric of spacetime, which we perceive as distortions in distances and in the flow of time. Take a few beads and place them neatly along what appears to be a circle. Don't tie those beads together, simply let them be "free." As a gravitational wave travels perpendicularly through the circle, it will stretch and squeeze the very geometry of that surface. The "free-falling" beads will follow the curved space, causing the position of the beads to appear distorted, as shown in figure 3.3. Like light, glight has two independent polarizations, at least according to general relativity.[6] They are known as "+" and "×" polarizations because of the way they distort the notion of space perpendicular to them as they evolve. These two polarizations exhibit the tidal force of gravity at its most basic level (at least until we account for quantum physics).

6. This is actually just an accident to living in a four-dimensional spacetime. If there were more spatial dimensions, glight would have more independent polarizations than light.

Note that even though the overall shape drawn by the beads changes as glight propagates through the page, the area encircled by the beads remains constant. Though you might appear taller and thinner for a while before becoming slightly shorter and wider as glight passes through you (if we were to ignore the other forces within your body that hold you together), rest assured that your overall volume will remain exactly the same throughout the process. This is, at least, according to general relativity with its + and × polarizations.

In theory, glight could carry up to six independent polarizations. In addition to the two mentioned above, there are three other potential types of polarization that stretch and expand the notion of space along the line of propagation of the wave while keeping the volume invariant. Then there is another polarization, dubbed the "breathing mode," which describes the expansion and contraction of the three-dimensional volume (see figure 3.4). Why did Einstein ignore these latter possibilities? In short, the equivalence principle and the powerful symmetry principles that follow from it, which are built into the very foundations of general relativity, forbid the presence of any of these additional four polarizations; their absence from our Universe can be regarded as a definitive prediction of Einstein's theory.[7]

---

7. If we consider gravity at the particle or quantum level (as we will in chapter 7), the principles and pillars on which general relativity rests can be seen to be by-products of more fundamental stability principles. Once we assume that glight, like light, is massless, then special relativity tells us that any other glight polarization would be so unstable that it would not only decay away, it would destroy the entire fabric of spacetime. Assuming that glight is massless guarantees that everything must be communicated equivalently with the same gravitational strength through these two polarizations, giving rise to the equivalence principle. The equivalence between gravitational and inertial mass is thus understood at the particle or quantum level as a consequence of glight's stability and lightness.

FIGURE 3.4. Breathing mode, also called conformal mode. This is an example of another polarization that gravitational waves could in principle also carry. This mode is absent in general relativity. *Source:* Claudia de Rham, "Massive Gravity," *Living Reviews in Relativity* 17, no. 1 (2014): 7.

To date, only the two polarizations of glight predicted by Einstein have ever been detected. But this won't stop scientists from looking for potential signs of others, the discovery of which would signal a departure from general relativity. These additional polarizations may not exist, or they may simply be rather shy and don't like to show themselves, or they may only appear under very special conditions.

Catching a Glimpse of Glight

Now that we've been introduced to glight, who wouldn't dream of sparking some of their own and feeling its gentle caress as they move through spacetime? The good news is that we all produce gravitational waves, everywhere, all the time. Being warm creatures, we give off a thermal "black-body" spectrum of radiation. This radiation is the emission of infrared light, outside of our visible range, at a wavelength of about 12 micrometers. At the same time, we emit a thermal spectrum of glight, although no cameras have yet been built that can detect it. We also emit glight every time we dance with our partner and every time our heart beats. In fact, electrons emit tiny gravitational waves as they spin around their nuclei at the core of every atom of every cell of our body. Everything, everywhere, all the time emits tiny gravitational waves that continuously communicate with the rest of our Universe. Sadly, most of them are undetectable. If we want to find glight of sufficient amplitude to be noticeable on Earth, we need to consider astronomical masses accelerated with strengths so intense that they would be unbearable to anyone in their close neighborhood.

Even an object as massive as the Sun cannot produce significant gravitational waves on its own.[8] However, when two stars are absorbed in an intimate dance, they can radiate enough energy into glight that the effect of this loss is noticeable. This is exactly what Russell Hulse and Joseph Taylor discovered in 1974 after observing two stars, each roughly the mass of the Sun, orbiting about each other just a few million kilometers apart. To us humans, those stars may still appear relatively far apart—about three times the distance between the Earth and the Moon. Exactly how intimate can you be if you are separated by millions of kilometers? While it's true that the two stars will not merge for at least a few hundred million years, they spin about each other in less than eight hours (compare this to the Moon, which takes a month to orbit the Earth). Their dazzling tango is maintained thanks to their intense gravitational attraction—so intense that they radiate glight with almost as much power (10^{24} watts) as all the light emitted by our own Sun (10^{26} watts). In fact, the Hulse–Taylor binary system radiates more power through the emission of glight than it does through the emission of light of any frequency (including invisible light). If we had evolved to be as sensitive to glight as we are to light, we might have heard the Hulse–Taylor system of stars through gravity before we saw it through light.

What makes the Hulse–Taylor system truly remarkable is the fact that one of the two stars is a *pulsar*. A pulsar is a special kind

---

8. The motion between the electrons and ions inside the Sun's core does produce thermal glight radiation but with such low intensity that we cannot detect it at this stage.

of star that, under the right conditions, emits thin beams of light from their magnetic poles. Like a lighthouse in the sky, a pulsar continuously emits light through its poles, but because it is rotating we only see a pulse of light when one of its poles momentarily aligns with our line of sight. The rhythm of the pulses we can observe from the pulsar is so precise that when involved in a spinning ballet with another close-by companion, we can infer the motion of the two stars with pinpoint accuracy. The measurements carried out by Hulse and Taylor were so precise that they were able to deduce that the distance between the two stars (the semi-major axis) decreases at a rate of three meters per year. Bearing in mind that the two stars are orbiting a few million kilometers apart, the precision of this measurement is like deducing that the distance between two people separated by one meter has shrunk by a nanometer—about the size of a small virus. The imputed rate at which the distance between the two stars is decreasing is in impeccable agreement with the change in orbit due to the emission of energy into gravitational waves predicted by general relativity. Just as the Sun gradually loses mass as it continuously emits light, giving it a finite life expectancy, any binary system of stars also gradually loses power through the emission of glight, eventually causing the two stars to merge.

This observation by Hulse and Taylor in 1974, just a few years after the discovery of the first pulsar, was the irrefutable albeit indirect proof that gravitational waves exist and are emitted by accelerating bodies.[9] (Forty years later, we would detect the

9. Dame Susan Jocelyn Bell Burnell was only twenty-four when, in 1967, she discovered the first pulsar together with Antony Hewish. Hundreds of astronomers have since followed in her footsteps, and thousands of pulsars have been discovered. Despite the fact that they do not emit a large amount of gravitational waves on their own, Bell's discovery has played a unique and essential role in our quest for gravitational waves and will continue to do so for centuries to come.

first direct glimpses of glight on Earth, confirming through other means what had been undeniable since 1974.) By now, thousands of pulsars have been detected throughout the sky. Some are in binary systems like the Hulse–Taylor one, and one is even in a triple system; but the majority of pulsars live a solitary, bachelor life far away from the rest of the stellar population. These single pulsars radiate even less glight than our own Sun. Nevertheless, even these bachelor pulsars can still play an important role in our search for glight. This is the ingenious idea behind the North American Nanohertz Observatory for Gravitational Waves (NANOGrav) and the overall International Pulsar Timing Array (PTA), which compares the precise timing at which light pulses from different pulsars are received on Earth. The timing of those pulses would follow a predictable pattern in a perfectly undisturbed Universe, but what happens when a gravitational wave passes by?

As gravitational waves ever so slightly distort the space and time that fill the gap between those pulsars and us, the path followed by the light the pulsars emit is in turn slightly perturbed. This leads to a correlated delay or advance in the time at which we receive the pulses from an array of pulsars, providing a smoking gun signature that a gravitational wave has passed between us or that we are embedded in a bath of glight continuously passing between the Earth and these pulsars. The power of such observations derives from the fact that they use the entire sky as our glight observatory. With the use of this enormous apparatus, the detection of a stochastic background of gravitational waves with wavelengths spanning across billions of kilometers was reported on June 29, 2023. This glight is being continuously emitted by a multitude of supermassive black holes billions of times the mass of our Sun, surrounding us in every direction [3].

Still using observations from PTAs, it may be possible in the future to detect gravitational waves with even longer wavelengths, which are expected to have been emitted during the early stages of our Universe, while it was still in its infancy. Detecting such glight would hence provide us with a glimpse into the birth of our Universe.

But pulsars are not the only tools that can allow us to observe gravitational waves. Although we typically associate the production of gravitational waves with the acceleration of very massive objects, in principle any object or substance produces gravitational waves, even something as light as light itself. Just by its very existence and by the energy it carries, light distorts the spacetime within which it radiates. The same is true for any wave or particle. In the future, we may reach a sensitivity where we could detect the gravitational waves emitted by the motion of the planets around our Sun or, for that matter, any exoplanet around any other star. If we succeed in doing so, we could use the loss of power through the emission of gravitational waves to provide us with new hints in the search for exoplanet systems.

For most of us, however, the amplitudes of the waves we generate are so small that they will never be detectable, not even by a detector the size of the Universe. Consider, for instance, two people involved in a rhythmic dance, half a meter apart, spinning around each other every 10 s. Their gracious sway would glow in more ways than I can describe, but the glight they emit will have an amplitude of about 10^{-43}. This amplitude is so ridiculously small that I cannot even imagine an experiment that would be able to detect this glight. For comparison, the wave depicted in figure 3.3 would have a rough amplitude of 0.5.

We can get a better idea of the typical amplitude of gravitational waves by looking at those emitted by the Earth–Moon ballet. These waves have an amplitude of about 10^{-24}. If two beads in empty space are located 1 m apart, with no other forces acting upon them (other than gravity), a gravitational wave of amplitude 10^{-24} passing by would displace the distance between those two beads by 10^{-24} m—that's about a millionth of the size of an atom, a ridiculously small displacement! For the Earth–Moon system itself, emitting such a gravitational wave would displace the distance between the Earth and the Moon by about 10^{-15} m, or about the size of an atom. While we currently measure the distance between the Earth and the Moon to within a millimeter accuracy, we would need to increase the precision of our measurements by about twelve orders of magnitude before being able to detect such a displacement. Nonetheless, such a feat has been accomplished: not by detecting the gravitational waves emitted by the Earth–Moon system, whose frequencies are too low for us to detect at present, but by detecting gravitational waves emitted by stars or black holes in the final few rounds of their dance, just as they are about to merge.

When two compact stars or black holes are about to merge, they spin around each other at close to the speed of light and emit gravitational waves with relatively large amplitude. As such events are relatively violent, you should take heed and keep your distance. Fortunately, the nearest mergers so far are dozens of megaparsecs away (about 10^{21} km). The glight that reaches us from these events was emitted more than a hundred million years ago, while the Earth was still deep in its Cretaceous geological era, dominated by dinosaurs and other now-extinct reptiles and ammonites. These waves have spread uniformly in every direction for hundreds of millions of years, gradually weakening along the way, reaching an amplitude of 10^{-20} or less by the time they hit the Earth. Gravitational waves of this magnitude pass though the Earth all the time, but in order for us to detect them, we need to use the most advanced state-of-the-art technology, and the waves themselves must be within a specific frequency, or color range.

To understand how we directly feel the caress of glight as it passes through the Earth, we need to go back to Michelson and Morley's infamous interferometer. It is perhaps ironic that interferometers, whose failure to detect the never-to-be luminiferous æther provided early evidence for special relativity, would also play a definitive role in the detection of spacetime fluctuations more than a century later, confirming a distinctive feature of general relativity.

When gravitational waves pass through the Earth, they deform distances along different directions (as shown in figure 3.3 by the circle of beads). To detect these deformations, we can use an interferometer similar to that designed by Michelson. It has two perpendicular, long vacuum cavities through which light propagates freely. These cavities are about 4 km long at both the Hanford (Washington) and the Livingston (Louisiana) sites of the Laser Interferometer Gravitational-Wave Observatory (LIGO) in the United States, and 3 km at the Virgo interferometer in Italy and at the Kamioka Gravitational Wave Detector (KAGRA) in Japan. As light reaches the end of the cavity, a suspended mirror reflects it back to where it came from, as can be seen in figure 3.5. The whole structure is attached to the Earth and is therefore bound with the very same gluing electroweak forces that connect all our atoms together. However, the mirrors at the end of the cavity are very delicately suspended so that they feel no forces from other neighboring atoms. That is the only way we can ensure the mirrors follow a straight path in our curved spacetime.

When a gravitational wave passes through the interferometer, it perturbs the spacetime in the experiment, and the mirrors are ever so slightly disturbed. As a result, the light from the

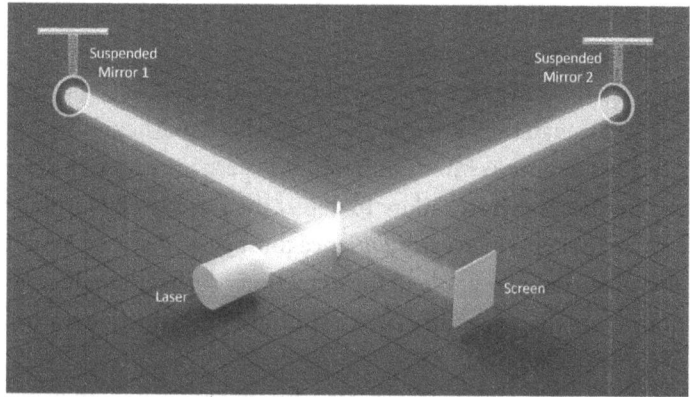

FIGURE 3.5. Sketch of the LIGO interferometer. *Source:* Adapted from screenshot of LIGO movie. All rights reserved. Courtesy Caltech/MIT/LIGO Laboratory.

lasers that shines through these cavities then reflects from the mirror at a slightly different spacetime location, which changes the path length of the arm of the detector. This small displacement is measured by comparing the relative phase of light sent along the two orthogonal arms of the interferometer.

Even with a cavity that is a few kilometers long, detecting the effect of a gravitational wave with an amplitude of 10^{-20} requires being able to measure a displacement on the order of the radius of a proton. Such displacements are so small that any local disturbance—of, say, a passing car or even a passing alligator near the Louisiana LIGO interferometer—could lead to similar effects. In addition to clever noise canceling designs, interferometers such as LIGO, Virgo, and KAGRA use additional mirrors in each beam to reflect light back and forth multiple times, increasing the effective arm length to up to 1,200 km for LIGO (200 km for Virgo). This amplifies the effect of the gravitational wave, causing a detectable phase shift between the two beams of light from each arm, which interfere when finally recombined.

When an oscillating gravitational wave passes through the interferometer, it creates a characteristic periodic fluctuation in the phase shift signal. The simultaneous detection of such a signal by two or more interferometers located at different locations on the Earth provides incontrovertible proof that a wave of gravitational nature just made its way through the Earth as it peacefully continues its journey to the end of the Universe.

I've been fascinated by glight ever since I was first introduced to it, long before I was even familiar with the details of general relativity. Yet I never imagined it would become an integral part of our scientific reality so quickly. Thanks to the recent detection of gravitational waves, we can already constrain some of glight's properties better than we can for light. Indeed, the glight signal we famously received on September 14, 2015, an event now known as GW150914, was not just a one frequency wave or a dull monochromatic gesture, it was a fully colored time-varying message with increasing frequency. And as this discrete message made its tenacious way through the Universe, passing through various cosmological structures before reaching us here on Earth some hundred million light years later, none of its colors were displaced—not a single one faded or drifted away. Such persistence and absolute perfection are unprecedented in the history of any other wave. No other wave has been proven to adhere to such a disciplined regimen, not even light itself. The very first direct detection of a gravitational wave pushed our understanding of gravity beyond what we could ever hope to do for light or any other fundamental wave or force.

Einstein Was Right! Or Was He?

To the surprise of many scientists, the first gleam of glight directly detected on Earth took place almost as soon as the two Livingston and Hanford LIGO detectors were first simultaneously switched on; in fact, they were still operating in engineering mode. However, the LIGO team spent several months checking and confirming their work before announcing their results on February 11, 2016. My colleagues and I were supposed to be sitting in a recruitment meeting at the time of the announcement, but we were in such a state of anticipation at the expected news that the meeting had to be cut short. We burst into the common room to celebrate the breakthrough with our students and postdocs. This was the start of a new era!

Waking up the next day, it was difficult to convince myself that it hadn't all been a fantastic dream. But these doubts were dispelled when I picked up my phone and saw discussions about the detection dominating every feed. Even arXiv, the online directory where we post new scientific publications before submitting them to a journal, was brimming with the wildest estimations, quickly constraining the most diverse set of gravitational models. Walking down the street, no matter where you were on the planet, you would see newspapers with an identical headline translated into the local language: "Einstein was right!"

Yes, Einstein was right about so many things. His theory of general relativity, the ludicrous idea that gravity could be the manifestation of the curvature of spacetime, was absolutely right. Einstein was right about Bose condensates. Einstein was right about random Brownian motion. Einstein was right about $E = mc^2$. Einstein was even right in saying that "joy in looking and comprehending is nature's most beautiful gift" [4]. But when it comes to the existence of gravitational waves, the claim is not as simple as that ubiquitous headline.

The concept of a gravitational wave was introduced by Einstein himself in 1916 as fluctuations or ripples in the structure of spacetime that could propagate almost free of hindrance. But Einstein was never pleased with the concept, which—according to him—would never be detectable. In June 1936, in collaboration with American-Israeli physicist Nathan Rosen, Einstein went a step farther in his dismissal of gravitational waves. According to their unpublished June 1936 paper "Do Gravitational Waves Exist? Answer: No!," not only would gravitational waves never be observable, but they weren't even physical. The first version of Einstein and Rosen's paper concluded that the presence of gravitational waves was no more than a mathematical artifact with no physical meaning; it would always be possible, they argued, to shift our notions of space and time in such a way that any gravitational wave would disappear, implying that gravitational waves were just an illusion. This was Einstein and Rosen's third paper together. The famous Einstein-Podolsky-Rosen paradox, dealing with the incompleteness for quantum mechanics, was published in 1935 under the equally provocative title "Can Quantum-Mechanical Description of Physical Reality Be Considered Complete?" That same year, with their article "The Particle Problem in the General Theory of Relativity," they discovered the first notion of wormholes in the interiors of black holes, also known as the Einstein–Rosen bridge [5, 6].

As their two previous articles had, their claim that gravitational waves were nonexistent was bound to shake the scientific community—or at least that was their expectation. In the days before scientists could post preprints online and receive feedback on their work before publication, one could never quite know how the scientific community would receive a finding. They submitted their manuscript directly to *Physical Review* for publication, the same journal that had published Einstein and Rosen's previous two papers in 1935. Until then, all of Einstein's manuscripts had been automatically accepted for publication. However, in 1936 a new selection process was introduced, one that now strikes fear in the heart of every researcher: peer review. In this case, however, peer review may well have provided Einstein with a lucky break [7].

While peer review is now ubiquitous—more than half of the manuscripts in my field are typically rejected by journals following the peer-review process, and almost all manuscripts require some editing before being accepted for publication—for Einstein and Rosen, the peer-review process was a novelty. When Einstein received the journal's response a short month later in July 1936 (nowadays, this process often takes several months), he was not impressed. The paper had not been rejected, but the authors were kindly encouraged to consider the comments raised by the referee.

Rather than subject himself to such scrutiny, Einstein decided to submit his paper instead to the *Journal of the Franklin Institute*. This incident possibly saved Einstein from the biggest embarrassment of his career. Unbeknownst to Einstein, the referee in question had been Howard Percy Robertson, a famous cosmologist known for his foundational understanding of our Universe. Luckily, Robertson was sufficiently familiar with general relativity to recognize flaws in Einstein and Rosen's argument. A few months later, before the article was printed in the *Journal of the Franklin Institute*, Einstein realized his mistakes and, after discussing the problem with Robertson, understood how to appreciate the physical nature of gravitational waves. This gave him the opportunity to correct the manuscript on November 13, 1936, before it was finally published in 1937 [8].

Thanks to Robertson, Einstein and Rosen's final published work, now titled "On Gravitational Waves," was correct. Einstein was indeed right after all. But this episode makes clear that detecting gravitational waves—an idea with which Einstein himself struggled!—is not about proving one scientist, or even one theory, right or wrong (or proving that it is in a superposed quantum state of right and wrong). Rather than providing closure as we end our story, the detection of glight is merely the start of ours. It is about contemplating a glimpse of nature in its most ancient and fundamental form and marveling at the journey it has taken through the Universe to reach us at this precise instant in space and time. Its journey does not end with us—though if we manage to decipher its message, we might unlock some of the Universe's most confidential secrets. It will travel onward, falling through space and time, as we fall along our own paths and lives. Our understanding may be incomplete, but that does not mean we cease in our quest to understand. To truly appreciate the Universe's wealth of mysteries, we must fasten our seat belts and dive into the fall.

CHAPTER 4

Predicting Our Own Fall

On April 12, 1961, Yuri Gagarin became the first man in space, ushering in a new era for humanity. Among 154 pilots, the Soviet cosmonaut was selected not only for his abilities, personality, fitness, and health, but also for his size. With a height limit of 1.70 meters (~5.6 feet), the selection criteria for the Soviet space program would have ruled me out from the start. After just over a year of training, Gagarin uttered his famous departing words—"Off we go! Goodbye, until soon, dear friends"—before being launched into space for a whopping 1 hour and 48 minutes. He was followed a month later, on May 5, by American astronaut Alan Shepard. Selected by NASA in 1959, Shepard was originally slated to take off in April 1960. Were it not for a series of post-ponements, he would have been the first man in space rather than "merely" the first American.

Since the 1960s, global space exploration has been expanding at an accelerated rate, much like our own Universe. Founded in 1975, the European Space Agency (ESA) selected its first astro-nauts in 1978. By that time, orbital space stations were already in their second generation. Soviet cosmonauts and American astronauts would team up and live together onboard the sta-tions for several months, carrying out scientific experiments and laying the groundwork for future space exploration. Three astronauts were chosen from ESA's first selection round in 1978: Ulf Merbold from Germany, Wubbo Ockels from the Netherlands, and Swiss astrophysicist Claude Nicollier. Five years later, Merbold became the first ESA astronaut (and the second European) to fly in orbit, as well as the first payload specialist to work on Spacelab-1, the first space laboratory. In 1985, Ockels became the second ESA astronaut to fly. With a PhD in physics, Ockels conducted extensive scientific experiments in space, from biology to material sciences.

Last but certainly not least was Nicollier. Although he had to wait until 1992 for his first space flight, his contributions to space exploration, and to science in general, are truly out of this world. In fact, as a child, I plucked up the courage to write a letter to my all-time hero about my desire to become an astronaut. To my astonishment, he took the time to respond to an insignificant ten-year-old living on the other side of the planet. His answer was imbued with such humanity and mindfulness that it still resonates with me decades later.

In 1993, Nicollier joined the STS-61 mission to fix the Hubble Space Telescope. The Hubble Space Telescope had been launched three years earlier, but once it was in space the mission scientists realized that its mirror was flawed. The mirror had what is known as spherical aberration (you can think of it as making the telescope slightly myopic), which meant pictures taken by this groundbreaking telescope were out of focus. The first servicing mission (STS-61) managed to correct the mirror with spectacles that worked much like the corrective lenses that many of us wear. Unlike a routine visit to the optometrist, however, STS-61 was one of the most challenging and complex missions carried out by NASA to date. Upon completion, it was declared a complete success; since then, the

Hubble Space Telescope has been uncovering new depths of the Universe.

Until the arrival of the James Webb Space Telescope in 2021, the Hubble delivered the most spectacular high-resolution pictures of astrophysical objects that humans have ever seen. Still today, thirty years later, images of Cepheids (pulsating stars) and supernovae (explosions of stars) from faraway galaxies taken by the Hubble Space Telescope continue to provide us with an ever more precise understanding of the origin, age, and evolution of the Universe. None of these insights would have been possible without this mission of incredible astronauts—including ESA's Claude Nicollier—who helped us, in their own way, reshape our understanding of the Universe.

In 1992, ESA undertook its second astronaut selection round. More than 22,000 applications were received, fifty-nine of which were selected by their own member countries for consideration by ESA. Ultimately, six astronauts were chosen, including the first ESA woman astronaut, Marianne Merchez from Belgium. Sadly, Merchez resigned a few years later, and we had to wait until 2001 for ESA's first female astronaut, Claudie Haigneré from France, to fly into space, though British cosmonaut and chemist Helen Sharman had flown with a Soviet team a decade earlier.

After years of preparation, my own attempt for the stars began on April 10, 2008, when ESA finally announced its much anticipated third selection round. With calls occurring every fifteen or so years, it was unlikely that I would get another chance. Of course, the same was true for an entire generation of eager scientists and pilots who had been waiting for the call for years. As soon as the selection process was announced, more than ten thousand applicants officially registered.

The first round of the selection process involved an online screening. That first step wasn't a simple tryout in my mind. For years, the dream of being an astronaut had shaped every single one of my days. In addition to possessing a science degree or a pilot's license, the application required a medical certificate and, most importantly, a letter describing one's motivation for wanting to become an astronaut. By that time, my desire to venture into space had been driving me for close to twenty years, but it still took me weeks to find the words to fully express this motivation and formulate how I saw myself contributing to the future of space exploration as a physicist obsessed with gravity. Of the registered applicants, only 8,413 candidates completed the first screening stage. I was one small step closer to space but still had many giant leaps left to go.

Out of these 8,413 candidates, 918 were selected and invited to Hamburg for the first round of psychological tests that summer. I was among them. At that stage, the dream of more than 90% of the registered applicants had already been cut short, but for me the challenge was just getting started. I had just arrived in Stockholm to lecture at a summer school on the cosmological constant problem. PhD students from all over Europe, as well as the People's Republic of China, the United States, and Iran, had traveled to attend the school, but I had to respectfully excuse myself for a day as I quickly popped over to Hamburg for the tests. Arriving at the hotel the night before my interview, I sat at one of the tables on the outside patio and heard a familiar voice. Having just flown from Canada, where I was living at the time, to Sweden then to Hamburg for the day, I was sure that jetlag was simply catching up with me. Imagine my surprise when I noticed that the person sitting next to me was Frederica, a friend I had not seen in eight years. Frederica and I had been roommates at the California Institute of Technology while interning at NASA's Jet Propulsion Laboratory in 2000, so perhaps it is unsurprising that our paths would cross again as we each pursued our passion for space.

The day that followed was no picnic. Each of the applicants in our group was placed in front of a computer and presented with a range of tests, including logic puzzles, English proficiency exams, scientific quizzes, and behavioral tests, to name just a few. As stress and fatigue built up throughout the day, the atmosphere surprisingly became more relaxed and friendly. Sweating our way through examinations in this stifling computer room, we realized that we were all in this together. Perhaps this had been ESA's intention. By the end of the day, we had formed an indescribable bond, and many of us stayed in touch during the nerve-racking months that followed.

One morning, a few weeks after the exams, the rejection messages started falling like rain. Frederica hadn't made it to the next stage. Neither had about forty other members of our group. I nervously refreshed my inbox throughout the morning, hoping that no news meant good news. After the dust had settled, 192 candidates from the first round of psychological testing had made the cut. To my disbelief, I was among the 2% of applicants who were invited to Cologne for a second round of assessments, which would include team-based stress tests and interviews with ESA members and astronauts.

On October 13, 2008, I presented myself at the European Astronaut Centre in Köln (Cologne), together with five other candidates, all men. The day was jam-packed with personal highlights, including the opportunity to meet astronaut Gerhard Thiele and to tour the astronaut-training facilities. The part of the tour I remember most vividly was the Neutral Buoyancy Facility, a huge, ten-meter-deep swimming pool designed to familiarize astronauts with the sensation of weightlessness in preparation for zero-gravity missions. As a certified dive-master, I was no stranger to the sensation, and I couldn't help imagining what it would be like to dive in this incredible glass-enclosed facility while working on space mission equipment. Somehow, that vision helped steel my nerves for the day ahead.

And what a day it was! We were subjected to a battery of tests, most of which were designed to assess our team-based abilities in stressful situations. The six of us began the day by being sent on a pretend rescue expedition through the "jungle"—we were asked to plan the risky mission knowing resources were limited, night was coming, and, if we weren't careful, some of us may not make it back. While we were brainstorming, members of the ESA team, including various psychologists and former and active astronauts, kept meticulous notes on our every move. Whether or not our fictitious rescue mission was ultimately successful was likely irrelevant—what mattered (or so I thought) was how we worked together as a team. Following our jungle adventure, we were paired up as pretend air traffic controllers in an over-crowded airport, tasked with landing several planes running short of fuel. The twist was that each of us only had access to half of the information, so we had to engage in precise communication and collaboration with our partner to complete the mission. As I was waiting for my turn, I remember one team storming out of the room, each resentful of how the other had handled the situation. It was then clear that my partner and I would either make it together or not at all. That day was simply not about us as individuals.

About eighty candidates were expected to be selected from this second round, but when the results were released in December, only forty-two had made the cut—my air traffic controller partner and I were both among them. None of the others from my group in Köln had made it. By that stage, more than 99.5% of the registered applicants had been rejected. I only needed to survive a little longer.

In early January 2009, I escaped the harsh Canadian winter for an interview for an assistant professor fellowship in Switzerland (a fallback plan in case my astronaut dreams were dashed). A few days later, on January 25, I set off for the third round of assessments: a week-long medical exam in Toulouse. I was joined by six other candidates (again all men) from the United Kingdom, Sweden, Germany, Finland, and Switzerland. We quickly called ourselves "The 7 Orbiters." The intimate medical screening left no stone unturned: I underwent X-rays, CAT scans, ultrasounds, needles, and a colonoscopy for good measure. They took stock of my blood, bones, eyes, lungs, heart, brain, and other organs. If anything was even remotely wrong with my body, they would find it.

At the end of the week, I had a debriefing with the doctor in charge. My results were excellent, he said, they were just waiting on the tuberculosis (TB) test results. By that point, I was laughing with him, certain that I was in the clear. Surely, if I'd had TB, I would have known. TB is accompanied by severe coughing spells, and I had been vaccinated for it as a kid. I'd also been tested for it on multiple occasions, and the results had always come back negative.

It turns out the joke was on me. The new and more accurate QuantiFERON TB test had only been recently introduced. I was about to turn off my phone before boarding my flight back to Canada when I noticed a message with the result: I was positive for latent tuberculosis, a little souvenir I must have unknowingly brought back with me from my time in Madagascar. I stepped into the plane, knowing that at that moment, my dream had evaporated into thin air.

Gravity at Its Limit

Would I still have trained for years and undergone all those tests if I had known from the beginning that I was fated to fail? I'd like to think so. Shooting for the stars is certainly a dream doomed to fail for most people, but I've never met anyone who regretted trying. The excitement of being one of the final forty-two candidates is an experience I wouldn't trade for the world. More importantly, I might never have learned to scuba dive or fly a plane, to dance with gravity in an intimate way that few people ever experience. Above all, facing failure is an essential part of the human experience, one that made me stronger and prepared me for a life of scientific research, where facing failure is an everyday occurrence.

Despite all the successes of general relativity, perhaps its most distinctive quality is that it predicts its own downfall. This is the most elegant feature a theory could ever enjoy, and one that has reshaped the way we now think about the laws of nature. Before the formulation of general relativity, the laws of nature were typically used to describe the world that surrounds us without offering a clear sense of when those laws would fail to provide an appropriate description. For example, Newton's laws of gravitation, despite their well-known problems, were believed to provide the correct description of the gravitational force in every situation. This belief was so firm that when scientists realized that the planet Mercury was not following quite the same orbit as predicted from Newton's laws, even accounting for the pull from the other known planets and other objects in the solar system, the only logical conclusion was to postulate the existence of a new, undiscovered planet.

The deviations from Newton's predictions were tiny—on the order of 0.000008%, ten times smaller than the effect that the other known planets in the solar system have on Mercury's trajectory. In 1859, French astronomer and mathematician Urbain Le Verrier was able to explain these deviations by postulating the existence of Vulcan, a small planet whose mass and location were predicted with astounding precision. And, indeed, on March 26, 1859, French medical doctor Edmond Modeste Lescarbault, a passionate amateur astronomer, claimed to have observed a dark spot transiting in front of the Sun exactly matching Le Verrier's predictions and proving the existence of the new planet. While he struggled to persuade the rest of the community, Le Verrier remained convinced of Vulcan's existence until the day he died. After all, this was the only way to reconcile well-known observations with the even better-known Newtonian theory of gravity.

We had to wait fifty-six years until, with the formulation of Einstein's theory of general relativity, scientists were ready to accept that the motion of Mercury was explained not by a new planet, nor other type of mass or matter, but rather by gravity itself. In other words, Newton's theory of gravity, while providing an excellent and accurate description for all previously known systems, had failed. When Newton formulated his laws of gravitation in 1687, there was no way to predict that, two centuries later, observations would reveal that small corrections would be needed to make the theory work—Newton's laws were believed to be the ultimate description of gravity until their connection with reality simply no longer made sense. General relativity, on the other hand, establishes realistic expectations from the start. The fact that general relativity will eventually fail to describe the fabric of reality has been taken for granted from the outset. The question that drives most theoretical physicists today is not *if* and *why* general relativity fails, but rather *what* comes next.

Singularities

We already know from Newtonian gravity that gravitational attraction increases as we move closer to a massive object and decreases as we move farther away. According to Newton's inverse square law (see chapter 1), as two objects move apart the gravitational attraction they exert on each other dims out as the square of the distance between them. This is why we don't spend our days worrying about being sucked into the gravitational belly of Sagittarius A*, the gargantuan black hole located at the center of our galaxy. There is no denying the overwhelming observational evidence that confirms its existence. And yet, we don't need to be too worried about it. Fortunately, or perhaps unfortunately, we have far more pressing issues to deal with on Earth these days. However, if we ever took a trip to outer space and found ourselves in the vicinity of an astrophysical black hole, our priorities might shift a little.

Taking Newton's inverse square law seriously at its other limit (i.e., when two objects are very close together), we would infer that as the distance between two masses vanishes, the gravitational force between them becomes infinite. This is what is known as a *singularity*.[1] There is a level of uneasiness, maybe even dizziness, associated with the idea of infinity—for millennia, humans have had difficulty making peace with the idea. And when it comes to singularities, nature faces the same problem. Luckily, the infinite force that emerges when the distance between two objects vanishes in Newtonian gravity should not be taken too seriously. We all know that it is simply impossible to localize a mass at a single point, as every object that has mass will also have

---

1. Throughout this book, a singularity will designate a point in spacetime where a quantity that can be measured, felt, or observed becomes infinite.

a width. If you bring two balls closer and closer together, eventually they will touch and cannot be brought any closer together. In Newtonian gravity, once the width of an object is taken into account, there can never be a real singularity. This is precisely where the black holes predicted by general relativity differ.

The laws of general relativity are governed by Einstein's equations, which we alluded to in chapter 3. Without going into too much detail, these equations tell us how matter curves spacetime and how this spacetime curvature affects anything living in it. As previously discussed, in most situations, these equations are impossible to solve exactly. Instead, one typically relies on simplifications giving rise to approximate solutions. These approximations provide a fair insight without being entirely correct or exact. In fact, there are only a handful of situations where we have access to exact solutions for Einstein's equations of general relativity. Black hole solutions are some of them: they describe bodies where gravity wins out over all other forces, creating a gravitational attraction so strong that even light cannot escape. If we were to take general relativity seriously at all scales, then black holes would be exact solutions of Einstein's equations, with no approximations being made whatsoever.

We typically think of curved spacetime as being "curved" by something—for instance, a star, a planet, or even a speck of dust. In the trampoline analogy we discussed in chapter 2, the ball placed on the trampoline caused its surface to bend. What is especially peculiar about black holes is that their existence does not require *anything* to be present; these exact solutions are "in the vacuum." Since there is nothing there, it means that there is nothing "stopping" anything from reaching the center of a black hole. Nothing can keep us from falling into its center, experiencing a point where the curvature of spacetime really is infinite. In other words, if we take general relativity seriously as a complete description of the Universe, then singularities are part of our reality.

This situation seemed so ludicrous that for many years after the development of general relativity few people took the idea of black holes seriously. In 1939, Einstein published a paper in *Annals of Mathematics* arguing that "Schwarzschild singularities" (i.e., what we now know as black hole solutions) "do not exist in physical reality" [9]. Yet in a remarkable set of papers between 1965 and 1970, Roger Penrose and Stephen Hawking proved several theorems that led to the conclusion that singularities were an unavoidable prediction of general relativity [10–13]. Far from being obscure solutions disconnected from reality, black hole solutions were recognized as central to general relativity, and as the inevitable end state of the gravitational collapse of a star (and potentially of our Universe).

To summarize: on the one hand, black holes with singularities at their centers are inevitable in general relativity; on the other hand, the existence of a singularity makes it impossible to describe our Universe at these points. How can we possibly make sense of the fabric of spacetime if the curvature—this thread connecting various points—can be pulled past its breaking point? The existence of singularities inside black holes would be so devastating to the fabric of our reality that the only logical explanation is that general relativity cannot be entirely correct. As we get closer to the center of the black hole, before we reach the singularity, general relativity breaks down and stops providing an adequate description of reality. A better description of gravity must take over. Penrose and Hawking's singularity theorems lead us to an unavoidable conclusion: if singularities are a prediction of general relativity but we can no longer make sense of the fabric of spacetime at those singularities, then general relativity has failed us there. General relativity predicts its own fall.

When we combine general relativity together with the laws of quantum mechanics, we can be even more precise. We know that general relativity cannot be trusted when the curvature scale reaches the Planck energy scale or possibly even before that. We understand that as we approach regions in spacetime when the curvature gets anywhere close to this Planck energy scale, a plethora of what we call quantum corrections comes into play and overpowers the predictions made by general relativity along the way. To appreciate why this happens, we must first discuss the central role played by the Planck energy scale in describing the fabric of reality.

Embracing the Planck Scale

In his 1687 law of universal gravitation, Newton needed to include a constant, G_N, that dictates the strength of the gravitational force that massive objects have on each other. This Newton constant, or gravitational coupling constant, cannot be inferred by first principles and must be determined empirically. It is about 6.67×10^{-11} m^3 kg^{-1} s^{-2}. To translate this mathematical jargon into more meaningful terms, if two astronauts were floating one meter apart in outer space, far away from anything else, the gravitational coupling constant dictates that they would exert a gravitational pull on each other of about 10^{-8} m/s^2. If they began at rest, after a few milliseconds they would be drawn toward each other at a speed comparable to that of your typical garden snail. So described, gravity seems relatively benign. But as they get closer, the gravitational pull becomes stronger, and the astronauts would fall into each other's arms in less than four hours. On Earth, other effects often come in the way, but in the middle of empty space, you can always count on gravitational attraction to bring you closer to the next massive object, and the rate at which this occurs is governed by this gravitational coupling constant.

Einstein's theory of general relativity goes beyond Newtonian gravity, but it still makes use of that very same constant. According to general relativity, our two astronauts would experience a similar outcome, at the same rate governed by that same constant. But in general relativity, this constant also dictates the extent to which matter, energy, and anything else living in our Universe curves the structure of spacetime and, in return, how much this curvature affects everything living on it.

General relativity is uncontestably the greatest of Einstein's many breakthroughs, but it is not the one with the most pop culture cachet. Most people will be far more familiar with his simple yet profound formula: $E = mc^2$. Einstein developed the famous equation while deriving his theory of special relativity, which paved the way to general relativity. According to this equation, a particle at rest carries an energy that is proportional to its inertial mass m and to the square of the speed of light c.[2] Through this formula, we see that the concept of mass is inextricably linked to that of energy. Almost every unit in physics can be converted back into energy in some form or another. Take the notion of time: if you have 10 minutes to walk 100 meters, you can do so at a gentle pleasant pace and not feel tired at all, but if you only have 20 seconds, you will likely be panting when you reach the finish line. A decrease in time is synonymous with an increase in energy.

Quantum theory solidifies this connection. It implies a duality between waves and particles, which I will discuss in chapter 7.

---

2. A little, often underappreciated fact is that this formula is actually only valid for a particle at rest. For a moving particle, its speed or momentum has to enter into the picture as well, leading to a more complicated formula. This is why a photon, which has no mass, can still carry energy—it is always on the move.

This duality translates into a relation between the wavelength λ of the wave, which measures the distance between two peaks of oscillation, and the momentum p of a particle. According to the de Broglie relation, $p = h/\lambda$, where h is Planck's constant. Embracing a quantum perspective, Planck's constant h thus allows us to convert every distance into momenta and vice versa. Together with the input of special relativity in the form of the speed of light c, we can use Planck's constant to convert time into energy.

Whether it is mass, time, or even distance itself, everything can be related back to an energy scale by using the two elementary constants of nature: h and c. Planck himself had realized as much already in 1899, in the same paper where he introduced his constant, a concept that turned out to be instrumental for quantum physics. By converting the *gravitational coupling constant*, G_N, into units of energy, he was able to derive what we now call the *Planck energy scale*, M_{Pl}. The Planck energy scale is roughly 100 kWh, about the amount of energy an average household in the United Kingdom would consume in ten days. Today, we understand that the Planck energy scale is the energy scale at which quantum gravity effects become important, which is why Planck's constant is crucial to its definition.

Fundamentally, we can and should treat gravity as a quantum phenomenon, just like light (here again, we will see in chapter 7 why this is the case). Below the Planck energy scale, accounting for the quantum nature of gravity is relatively simple, as the quantum corrections that describe how quantum effects differ from those of classical physics are small and remain under control. At low energies, which is where we expect general relativity to be valid, gravity appears as a classical phenomenon. However, things change dramatically when considering situations where the spacetime curvature becomes large and reaches the

Planck energy scale.[3] In such cases, the quantum corrections overpower the classical contributions and become intractable. With the tools we currently have at our disposal, there is no determining what the precise outcome will be, but there is no expectation that it will follow the same pattern as general relativity's classical predictions.

One hundred kWh may not appear to be very large, especially when you're trying to keep your house warm in the middle of the winter. On the other hand, from the point of view of the other fundamental forces of nature that govern the laws of physics— the electroweak and strong forces that explain everything from chemistry to the internal structure of atoms and their nuclei— this scale is absolutely enormous. It is about twenty-two orders of magnitude larger than the energy carried by an electron at rest. To confine such a high level of energy within the size of a proton, for instance, would require a particle collider fifteen orders of magnitude more powerful than the current state of the art, the Large Hadron Collider (LHC) at CERN (Conseil Européen pour la Recherche Nucléaire, or European Council for Nuclear Research). Given that it takes decades to ramp up the energy levels at particle accelerometers by just one or two orders of magnitude, and that each successive order of magnitude becomes exponentially more difficult, you can only imagine what it would take for a particle accelerator to reach the Planck energy scale. I am not going out on a limb by predicting that there will never be a Planck-scale particle accelerator on Earth. If there were though, within the current laws of physics there is no predicting what the outcome of such an experiment would be or whether the Earth itself would even survive it. We

3. For comparison, the surface of a sphere has a curvature on the order of the Planck scale if the sphere has a radius of about 10^{-33} cm (the Planck length).

only know that the current laws of physics we use to describe reality would no longer apply.

In spirit, this is not so dissimilar to what happens to Newtonian mechanics as we approach the speed of light. Newtonian mechanics is perfectly acceptable and predictive at sufficiently slow speeds, but as we accelerate to speeds closer to that of light, Newton's theory fails, and corrections from special relativity become important. The key difference with general relativity is that we already know from the outset that it will break down as we approach a curvature close to the Planck scale, even though we have no idea of what actually happens when we get into that regime. While we are still awaiting *the* theory of everything that may provide a clearer picture of what happens at that point, general relativity has already provided us more than we could have hoped for: the prediction that it will fail at or below the Planck energy scale. General relativity has predicted its own downfall—and by doing so, opens up an opportunity to move beyond the limits of our current understanding.

Our Journey inside a Black Hole

On the surface of the Earth, the spacetime curvature is about forty-five orders of magnitude lower than the Planck energy scale. It never gets much higher anywhere in our solar system, or at least not right now. But neither the Earth, nor the Sun, nor any other planet in the solar system will be around forever. Our solar system was born from the leftover gas and dust expelled from the explosion of a star, or supernova, some 4.5 billion years ago, and pulled together thanks to the gravitational force—a gravitational force so intense that it was, by itself, sufficient to ignite the gas at the core of the Sun and start the process of stellar combustion. If day after day, year after year, our Sun appears to be unchanging, this is only because the expanding pressure from the gas at the Sun's core and the collapsing pressure of the gravitational force are in nearly perfect equilibrium.

But a time will come, in about five billion years, when the Sun will start running out of fuel and turn into a red giant, swallowing the Earth along the way. After about another billion years or so, the pressure of the gas will no longer be sufficient to counteract the strong gravitational pull at its center. At this point, the Sun will most likely shed its outer layers into a planetary nebula, feeding the birth of new stellar systems, while the Sun's core, no longer able to fight gravity, will collapse in on itself. We predict that a large portion of the solar mass will then be confined to a region not much larger than the size of the Earth—our Sun will become a remnant white dwarf faintly glowing from the leftover heat. This will be the most dramatic event in the history of our solar system, and, one way or another, humanity will be long gone by then. From the perspective of general relativity, however, this will still be considered perfectly smooth sailing: the spacetime curvature of our white dwarf Sun will remain some thirty to thirty-five orders of magnitude away from the Planck energy scale, keeping it blissfully in a regime where we can fully trust its predictions.

Our Sun, though, is not particularly massive, at least not when compared with the other stars in the Universe. For comparison, consider Orion. Even in London, where I live at the moment and where the stars never appear particularly bright, the constellation is one of the few that stands out. Rigel, its brightest star, is a blue supergiant about twenty times the mass of the Sun. At the moment, the curvature at its surface is still very weak, close to forty-eight orders of magnitude below the Planck scale and well inside general relativity's comfort zone. The laws of general relativity are well tested on spacetimes with such low curvatures, and they agree perfectly with our empirical observations. With such a large mass, however, the gravitational pull will be so strong at the end of Rigel's life that most of its remaining mass will be confined to a region smaller than about 60 km wide. At the surface of this incredibly small yet unbelievably dense core, the curvature would still be some thirty-nine orders of magnitude away from the Planck scale, and the laws of general relativity would still hold. Yet they would imply that the gravitational pull would be so strong that as one approaches Rigel, it would become harder and harder to pull away from it, up to the point when you reach a horizon, or surface of no return. From that horizon nothing, not even light itself, would ever escape. Like the current pulling a swimmer toward the edge of a waterfall, which eventually becomes so strong that it is impossible to swim fast enough to avoid being dragged down the waterfall by the flow, the gravitational pull on that surface is so strong that once you cross it you can never return. From Rigel, and any other sufficiently massive star at the end of its life, a black hole will be born.

The idea that the Universe is filled with black holes—regions of spacetime that are so opaque that no one and nothing could ever peek inside without being stretched into pieces and swallowed—may be quite discomforting at first, so it is perhaps not entirely surprising that the scientific community took its time to embrace the idea. By now, however, the existence of black holes is more than well-accepted—it is understood to be unavoidable. As noted above, this was first realized by Roger Penrose and Stephen Hawking in the 1960s, a breakthrough recognized with the 2020 Nobel Prize in Physics, half of which was dedicated to Penrose "for the discovery that black hole formation is a robust prediction of the general theory of relativity." (Hawking, sadly, had passed away just two years earlier and so was ineligible for the prize.) Since the 1990s, thanks to the W. M. Keck Observatory in Hawaii,[4] we have been able to observe the effects that the supermassive black hole living at the center of our galaxy has on neighboring stars. Over the past three decades, the Keck telescopes have tracked the motion of stars at the center of our galaxy, revealing that they are all orbiting around an invisible point so massive and yet so compact that it is most likely a black hole. This discovery garnered the other half of the 2020 Nobel Prize in Physics, which was awarded to Andrea Ghez and Reinhard Genzel for the discovery of a supermassive compact object at the center of our galaxy, making Ghez just the fourth woman to win the physics Nobel Prize in its 120-year history.

During the past decade, we have also been blessed with a multitude of independent evidence for its existence, including the first "image" of Sagittarius A* at the center of our galaxy, revealing glowing gas and bending light surrounding a dark region at the center of our very own galaxy. By their very nature, black holes cannot be photographed, at least not directly, because no light can ever leave their surface. Nevertheless, the Event Horizon Telescope (EHT) achieved the almost impossible in 2019 by capturing the shadow of the object at the center of galaxy Messier 87 and in 2022 that of Sagittarius A*. The EHT is a network of eight radio telescopes working together across the planet. Light emitted from behind that object was observed either to bend around its surface or, if it dared to come too close to the object, to be captured by it, providing the closest picture there could ever be of a black hole—again, in impeccable

4. First designed in the 1970s, the Keck Observatory became fully operational in 1996. It includes two telescopes with what at the time were the largest mirrors in the world.

agreement with predictions from general relativity in a regime where it is safely under control.

Additional evidence was provided by the first direct detection of gravitational waves by the Laser Interferometer Gravitational-Wave Observatory (LIGO) in 2015 (as discussed in chapter 3), as the waveform detected precisely matched that predicted by general relativity during the merger of two black holes. Although we do not know exactly when the two objects initiated their complicit dance—it could have been millions or even billions of years ago—we do know their size: the gravitational wave signal they emitted in the last few milliseconds of their dance, right before they united into one, bore the imprint of their respective masses (about 30 and 35 solar masses). The mass of the combined object was about 62 solar masses, and an estimated 3 solar masses were radiated in the form of gravitational waves. What's more, the amplitude and frequency of the signal emitted are in perfect agreement with the predictions from general relativity of a black-hole merger.

Since we didn't work through the detailed derivation of Hawking and Penrose's pen-and-paper proof for the inevitable existence of black holes, you would be forgiven for remaining respectfully skeptical. And while the combined observational evidence collected by the Keck and Event Horizon Telescopes and by the LIGO-Virgo-KAGRA observatories leaves little room for doubt, perhaps an in-person visit is what would finally convince you of such an unthinkable phenomenon. But before you book your flight, make sure to read the fine print on your travel insurance policy: if you were to dive headfirst toward the black hole at the end of Rigel's life, the spacetime distortions would be so intense that as you approached its horizon you would age hundreds of times slower than the friends and families you left behind. From their perspective, it would appear as if you had stopped aging altogether as you crossed the horizon or surface of no return.

Throwing yourself into a black hole may seem to provide the ultimate antiaging cure, but it comes with a few less pleasant side effects. Together with the distortion of time comes an unavoidable distortion of space, which would squeeze your body horizontally while stretching it vertically. Young and thin, what's not to like? Well, the vertical stretching would eventually be so strong that the electroweak force that holds together the particles in your very atoms would no longer be sufficient to maintain you together, and your body would be ripped apart at the fundamental particle level.

During your rather unpleasant and ultimately fatal black hole expedition, rest assured that theoretical physicists armed with the laws of general relativity would maintain perfect confidence in your fate throughout your journey, even past the black hole horizon, or surface of no return. For a black hole of the mass of the Sun, the curvature at its horizon is still thirty-nine orders of magnitude away from the Planck scale—crossing it is not where general relativity fails. The real drama doesn't begin until you've entered the black hole and are deep within its core. The black hole does not simply swallow you up and stretch you out—it also pulls you inexorably to its center.[5] However, as you fall toward the center, the curvature rapidly ramps up, and that curvature eventually becomes uncomfortably close to the Planck scale.

Every notion of physics as we know it crumbles at this point. This is where we know general relativity will fail—the theory

5. This statement is a prediction of general relativity in a regime where the curvature remains very small compared to the Planck scale, easily thirty to forty orders of magnitude smaller, a regime where general relativity has been well tested and can be trusted.

itself tells us as much—without ever needing to observe any contradiction. Taking the theory at face value in the center of the black hole would suggest an infinite curvature, or a singularity. But as we have seen earlier, this concept is meaningless. What takes over and makes sense of reality in that region is not fully known, and we may never find out. But embracing the failure at the heart of general relativity is the only way for us to get closer to the truth.

Our Journey to the Beginning of Time

Black holes have fascinated scientists for nearly as long as the theory of general relativity itself. Despite being stationed at the Russian front during World War I, German physicist Karl Schwarzschild needed only twenty-seven days from the publication of Einstein's equations to derive the first exact black hole solutions. It took only twenty-seven days for general relativity to disclose its singularities and reveal a region of spacetime where our notion of reality simply falls apart. Yet black holes are not the only instances in the Universe where general relativity reveals its failure. To push general relativity to its limits, we must travel back in time.

Our Universe is about 13.8 billion years old—or at least this is how long it has been since the Big Bang. If we could travel back in time, into the history of our Universe, we would experience an environment that is much hotter and much denser than our present surroundings. Before the Universe was 380,000 or so years old, it was so dense that protons were unable to trap electrons around their nuclei, making it impossible for atoms to exist. The entire Universe was a dense, piping hot soup (4,000° Kelvin, close to 7,000° Fahrenheit) of fundamental particles that were constantly bouncing off of one another. With no stars yet formed, no structure, and not even the simplest atoms, there would be nothing to "see." At that time, the curvature of spacetime was about twenty orders of magnitude greater than what we experience today in the solar system, but still well below the Planck scale.

Going back even farther to just 10^{-12} seconds after the Big Bang, the Universe was so hot and dense that neither electrons, nor light itself, nor any of the other fundamental particles with which we are familiar would even exist in their present form. The spacetime curvature was then just a few orders of magnitude below the Planck scale, but still well within the safe region where the predictions made by general relativity can be trusted and predictions can be made. Rewinding even farther, before 10^{-12} seconds, the Universe was even denser, crashing any sense of reality we could imagine, and yet general relativity was still valid. We can go even farther back in time, to when a period of cosmic inflation is thought to have taken place. Current observations suggest that at around 10^{-33} seconds after the Big Bang, as the curvature was still below the Planck scale, the Universe must have undergone a sudden phase of rapid acceleration, expanding its size by at least fifty orders of magnitude in just a fraction of a second. During this phase of cosmic inflation, the quantum fluctuations of the spacetime curvature, although smaller than microscopic at that time, grew to cosmological sizes, seeding the fluctuations in matter distribution that later formed the clusters of galaxies and the structure of the Universe we know today.

Heading farther into the past, the curvature of spacetime inexorably increases until, roughly 10^{-43} seconds after the Big Bang, it becomes comparable to the Planck scale, the scale that dictates everything we know about gravity, even its failure. Just as general relativity inexorably breaks down as one moves sufficiently close to the center of a black hole, the same happens as we rewind back to the birth of our own Universe. This is no coincidence—as Stephen Hawking and Roger Penrose pointed out in the 1960s, the cosmic singularity at the origin of our Universe forms in a very similar way to the singularity at the center of black holes, and both are unavoidable in general relativity.

Let's pause for a moment to take it all in. The failure of general relativity as we approach the Big Bang is more than just a mathematical curiosity on a piece of paper. It challenges us on an existential level. It's one thing for us to grapple with an early Universe where the fundamental particles we are made of don't exist—that's not pleasant to consider, but we can deal with it. But the fact that we must dispense with the concepts of gravity, space, and even time—as we must do at the Big Bang—makes understanding our own origin so challenging and fascinating. The realization that general relativity fails at the Big Bang means that the origin of our Universe lies beyond our understanding of reality at the moment—and perhaps always will. It is not just that we lack the tools to explore our origin; we do not even have the proper language to formulate the questions we would like to ask. Without even a notion of time at our disposal, the very question of what happened *at* or *before* the Big Bang is nonsensical. This failure of general relativity may be regarded as one of its greatest disappointments, but in return it has provided us with the opportunity to embrace our lack of knowledge until new layers of reality emerge.

Nowadays, most physicists, and particularly theoretical physicists like myself, do not believe that there is an absolute, ground-level notion of truth that will describe the fabric of reality. Rather, we have adopted the perspective that nature is a superposition of layers that we are gradually uncovering, one after the other, digging ever deeper in our understanding of nature even as we recognize that our search will likely never end.

The deeper our understanding, the more aware we become of our ignorance and limitations, and the better we can appreciate how much more there is to discover. General relativity has painted a close-to-perfect picture of nature on our canvas, but it is still just a picture. Reality exists beyond it, and as a scientist I am compelled to find new ways to describe it as accurately and vividly as possible. With this in mind, a naive and yet natural question pops up: does general relativity fail only at large curvatures comparable to the Planck scale, or does it also fail at small, ridiculously tiny curvatures? This simple question has paved the way for thousands of investigations and changed the way we think about our Universe and gravity itself. This is also where, fresh off my own failure to reach space, I began in earnest my own pursuit of gravity. But to appreciate how and why this happened, we first need to take a further journey through the past and ultimate destiny of our Universe.

CHAPTER 5

Expanding into Nothingness

In the months that followed my tuberculosis screening results, I underwent an intense regimen of antibiotics. But the treatment, a precautionary measure more than anything else, would do nothing to change the diagnosis of latent tuberculosis, which would remain with me forever. There was no fighting or denying it. I had to face the fact that, no matter how much I trained or put my mind to it, floating in space and walking on the Moon would never be part of my life.

Rather than seeing it as the end of my adventure with gravity, I like to think of it as the opportunity for a new beginning. Although I would never be accelerated at 3 G's before encountering weightlessness, nor experience gravity in its most dramatic setting, the intimate dance I felt we shared would continue. As a scientist I could still experiment with gravity, and tease it in my own way. This would mean embracing even more grueling challenges: those of academia, where the persistence and motivation that had been fueling my astronaut dream would come in quite handy. Fortunately, I wouldn't be forced to confront those challenges alone. Throughout this new adventure, I could always rely on a sense of cohesion with other scientists, as I had with the other candidates during the astronaut selection process.

Theoretical physics is often portrayed as a field of solitary geniuses, with each person toiling away on their own until they stumble upon their eureka moment. In fact, if you search for "physicist," the first picture that pops up will invariably be one of Albert Einstein, alone, at the age of sixty-eight. There is no doubting Einstein's distinctive genius, and he was widely renowned by the time he had reached his seventh decade, but it may be refreshing to state the obvious: despite the way our community chooses to remember him, Einstein was neither old nor alone when he made the majority of his discoveries. He was only twenty-six in the "miraculous year" that he unveiled special relativity, Brownian motion, the famous equivalence between mass and energy, and the work that would lead to his Nobel Prize. And these breakthroughs would not have been possible without the contributions of numerous other amazing scientists, including James Clerk Maxwell, Max Planck, Ludwig Boltzmann, Heinrich Hertz, Hendrik Lorentz, and countless others. One of the most difficult and yet crucial skills required to survive as a theoretical physicist is the ability to work together, as part of a group effort. For some of my colleagues, their individual genius and ingenuity are likely enough to keep any risk of failure at bay, but that certainly was not the case for me, nor is it the case for most scientists.

The other crucial trait that any successful theoretical physicist must develop is the ability to persevere through long and often disappointing periods before progressing in their research or in their academic career. Every year, only a handful of faculty positions in high-energy theoretical physics are advertised around the world. These positions will attract hundreds of the brightest candidates, each of whom will have been the highest-performing student from their respective college. They will also have completed years of fruitful postdoctoral work. Securing one of these positions is a triumph. Securing one in a particular location on the planet is a tour de force. Securing one that fits your ideal personal and family needs is essentially a fantasy. This was the reality that Andrew and I faced as we navigated the complexities of academia. Andrew had had a fantastic career thus far, and soon enough offers were in his hands. This was not my case. I wasn't naive about the odds, but I knew my best chance at success involved moving steadily forward without letting myself be discouraged by all the accompanying fears and anxieties.

Fortunately, by that stage I had grown accustomed to persevering. Whether it was continuously joining horse-jumping competitions, only to be disqualified in the first obstacle, or having to prove myself every time I moved to a new place, restarting at the bottom of the ladder was nothing new. As a young adult, after eight wonderful years in Madagascar, I returned to Switzerland to study physics. The transition was bound to be easier this time, or so I thought, as I was moving back to somewhere I knew, to do something I loved. Ironically, I never felt more like an outsider than when I returned to what was supposed to be my "home country." It wasn't so much that I was attempting to cope with the cultural shock or that the ratio of women to men in the major was so lopsided (only around 10% of the students in first-year physics were women at the time, and it dropped sharply after that to a total absence of women professors or lecturers throughout my degree), it was that women were simply more expected to fail. And, indeed, the overall trend seemed to "confirm" this expectation: I recall raising my hand to answer a question posed by one of our professors in my first year and being turned down with the remark, "Ha, a woman who thinks she knows the answer. Surely that can't be right." He then urged one of my male peers to answer the question. In questioning my place in the field, even in his jocular manner, my professor was not alone. Only a few years ago, I was invited to give a series of lectures on dark energy for theoretical physics graduate students at a summer school in Italy. The students were just amazing, fully engaged and eager to understand. I remember one student in particular—he was so excited that he kept asking questions throughout my lectures. Reaching the end of my last lecture, we continued our discussion for quite a while until we both had to rush down for lunch before the cafeteria closed. Later that day, I spotted him having dinner on his own, so I went up to him to say hello. In response, he stared up at me in surprise with a puzzled look in his eyes, wondering why I was interrupting his dinner. After a long awkward silence, he finally asked me if I was the wife of a physicist attending the school and if I knew anything about physics myself. By that point, I had grown so accustomed to the idea that others would decide from the outset that I couldn't possibly have anything meaningful to say about physics that I just laughed as I responded, "No, what do I know about physics? Surely nothing," admitting that there is still so much more we need to uncover after all. By now I have lost count of the number of times I was told that persevering in physics was bound to be too challenging for me or the number of times I was also told that there was no point in attempting to carry out a career of my own when I could be so much more useful following my partner and supporting science in other ways. Worse, perhaps, was being repeatedly mistaken for the nanny, the helper, or the coffee maker (why else would I be there?), being asked to justify my presence, or simply being dismissed altogether.

While I would never endorse this behavior, I can say, with some level of amusement, that in the long run experiences like these simply made me impervious to the judgment of others, or even to my own poor performances. For me, the only viable option was to focus on moving forward while brushing off any negativity. Of course, there is a fine line between tenacity and stubbornness, and who knows how many times I crossed it. But I'm certain that my irrational stubbornness and naive optimism, both of which brought me to the precipice of becoming an astronaut, contributed to my success in academia. Persistence cannot itself guarantee that one's pursuit will have a happy ending, but it is a crucial part of any story that does. In fact, when it comes to perseverance, cohesion, and happy endings, the Universe may have the best story to tell.

In this chapter, we will explore the farthest reaches of the Universe and discuss the discovery of a new substance called "dark energy." Over the past two decades, much of my own research, as well as that of hundreds of other physicists, has been devoted to uncovering its nature. We will discover some of its intriguing properties and come to appreciate how it may be connected to the nature of gravity. For now, however, what fascinates me the most about this dark energy is that it instantiates how this most insignificant and negligible element could, through nothing other than sheer persistence in time and cohesion in space, ultimately end up being the most abundant and important substance in the entire Universe, the one that will eventually determine its fate.

The best mysteries are often those that end with a twist, where the real protagonist is not a single person but an entire undercover community you never suspected was working together, one you didn't even notice until the very end. In the twisted history of our Universe, dark energy is that agent. It is the character that has been present throughout the ages, lurking in the background, apparently so insignificant that its presence went unnoticed for billions of years. While everything else was busy pushing the limits of the expanding Universe, dark energy stayed put throughout the ages, as if part of the greatest collective effort in all of history. It never changed, never evolved, never made a sound, at least not until everything else had faded away (or perhaps had been converted into dark energy itself).

Today, the main actor in the story of our Universe, the one who will lead the grand finale, is neither the galaxies nor the rich clusters of dark matter scattered throughout empty space. It is dark energy, persevering through eternity and spreading constantly everywhere. Its behavior will ultimately decide the fate of our Universe and the destiny of space and time themselves. It is a tribute to what incredible persistence and cohesion can achieve.

A Mysterious Universe in Continuous Expansion

To learn more about our Universe, let us leave our planet for a moment and consider distances beyond our solar system, beyond our Milky Way galaxy, all the way to Andromeda, our neighboring galaxy. Even at this scale, which is relatively small in the grand scheme of things, there are already hints that there must be more to our Universe than meets the eye. At such distances, we see that stars orbiting around galaxies do so as if pulled by a gravitational mass far greater than what can be accounted for by the mass we see; the combined mass of all the atoms in the stars in the galaxies, all the planets orbiting them, all the comets and interstellar dust, even the neutrinos is simply not sufficient to explain the gravitational pull felt by stars orbiting about the galaxy. To account for their accelerated pull, all the galaxies and clusters of galaxies in the Universe must all be submerged in an invisible substance, which we fittingly refer to as "dark matter." The abundance of dark matter is more than astronomical. It turns out that all the stars and dust and other particles that we know to exist make up only a small fraction of the mass of our own galaxy. The rest must be accounted for by dark matter.

The exact nature of dark matter is unknown, though thousands of scientists are working hard to figure out what it is. Perhaps a small lump of dark matter is passing through your body right now, popping out and fidgeting in front of your nose as we speak. If so, you would feel nothing. We've never *seen* nor detected dark matter directly. So far, it appears to want nothing more than to be kept hidden from us, only communicating with us via gravity. The gravitational pull of such a small lump of dark matter would be zillions of times smaller than any other sort of disturbance you normally feel on Earth. In fact, it would be so small that even a single oxygen atom from the air would exert more pressure on you. Nevertheless, when zooming out to large distances, the effect of this as-yet-invisible matter is unmistakable. Its gravitational impact on stars, galaxies, superclusters of galaxies, and the Universe as a whole is not only prominent but indispensable.

There is now plenty of observational evidence that dark matter has been instrumental in the evolution of our Universe since its early toddler years, around the time it celebrated its adorable fifty-thousandth birthday. The gravitational pull of dark matter is responsible for planting the seeds of the first galaxies and for the birth of the first stars. We may not feel its gentle touch as we go about our daily routines on Earth, but its presence throughout the ages of the Universe has been vital to our existence.

Now that we've taken dark matter on board, let's continue our journey a little farther, zooming out past Andromeda to even greater distances, passing through our local supercluster of galaxies, known as the Virgo Supercluster, which spans hundreds of millions of light years. Curiously, those galaxies and clusters of galaxies are observed to be moving away from us in all directions. In fact, the farther away a galaxy appears to be from us, the faster it appears to want to get away.

If we did not have the benefit of thousands of years of accumulated wisdom, we would be tempted to conclude that our very own galaxy, the Milky Way, holds a special place in the center of the Universe, from which every other galaxy is attempting to flee. Yet just as we have each learned to face the fact that we do not occupy a special, central place on Earth, scientists have learned to accept that the Earth does not occupy a special place in the solar system, nor the solar system a special place in our galaxy. Wisdom tells us that the same should hold true for our galaxy. While putting ourselves at the center of the Universe may be a convenient explanation for some observations, it leads to the much more troubling question of why and how we came to be so privileged.

Instead, a much more credible explanation is that every point is receding away from every other point democratically. From the point of view of *any* galaxy in the Universe, everyone else always appears to be moving away. In other words, our Universe must be expanding. To be a bit more precise: if our observations reveal that the galaxies are all moving away from one another, it is because the space between any two points in the Universe is being stretched. To picture this, we can zoom back in on Earth for a moment, to one of my daughters' birthday parties. As I exhale with all the power of my lungs, the points on the surface of the balloon I am blowing up all miraculously drift apart, and the writing "Happy Birthday" grows larger and larger. Take any two points on the surface of the balloon, and you will notice them moving away from one another. Of course, our Universe is not the surface of a rubber balloon, so the analogy will only get you so far. Most notably, our Universe is not expanding *into* something else as the balloon is.

When the concept of *expansion* gets mentioned, it is difficult to resist the urge to picture the Universe with a finite and growing boundary. Our natural instinct is to wonder what the Universe could possibly be expanding into. Unfortunately, the only answer I can give you is the one that I am sure will satisfy you the least, but let me say it anyway: the Universe expands into *nothingness*. To be more accurate, the Universe does not expand *into* anything. It simply expands by itself within itself. In contrast to our inflating balloon or, say, a pipe leak that causes gas to spread into a room that exists in its own right, our expanding Universe is not spilling into another separate entity, nor even another dimension. It is the structure of space and time that stretches, a structure that has always existed, at least since the Big Bang. This structure or fabric of space and time is elastic and malleable. Just as it can pull us apart as we fall inside a black hole, it can also expand and cause anything and anyone embedded in it to drift apart.

Although we cannot say for sure, it is quite likely that the Universe is infinite in size. Because light and glight travel at finite speed and there has only been a finite amount of time since the Big Bang, we will only ever be able to observe a finite portion of this infinite region. Thus, our *observable Universe* is finite. But the *entire Universe* itself exists beyond what we can see and feel and smell and observe, beyond what we can see through light and hear through glight. In fact, throughout the cosmic expansion, the size of our observable Universe may not necessarily grow. If the expansion accelerates rapidly enough, our observable Universe would shrink in size because light and glight from far away would find it harder and harder to compete with this expansion.

Physical boundaries and inexorable endings form such an essential part of our life on Earth that the prospect that our Universe may be infinite in space and time can be unsettling. But the Universe does not care about satisfying our instincts; if it is infinite today, it is because it was "born" that way. The Universe would have been infinite in size from the moment it was conceived at the Big Bang (possibly even before) and may remain so for an infinite length of time. Being infinite both in space and time, it can grow, evolve, and blossom without the constraints of boundaries, deadlines, and other limitations to which we are so accustomed.

An Accelerated Expansion

There are now numerous independent probes that allow us to infer how the Universe has been evolving since its creation. One probe, first used by Edwin Hubble, are Cepheid variables, a type of pulsating star whose temperature and brightness change periodically. In 1908, Henrietta Swan Leavitt realized that the time period over which these Cepheids pulsate was linked to their absolute luminosity, meaning they could be used as *standard candles*. In astronomy, a standard candle is a body that emits light with known power or luminosity, which can be used to measure the distance to their host galaxy. In the case of Cepheids, by measuring their pulsating period, one can infer how bright they really are (their absolute luminosity). Then, by comparing their actual brightness with how bright they appear to us, we can infer how far away they are—this is what astronomers call the "luminosity distance." Today, one of the most powerful standard candles used by cosmologists are type IA supernovae.

A supernova is the spectacular explosion of a star that occurs when its internal pressure can no longer sustain its own weight. These cataclysmic events involve an emission of light that is billions of times brighter than the Sun and which peaks at a very specific frequency. If the galaxy hosting this exploding star is moving with respect to us, the frequency of this peak will appear shifted due to the Doppler effect. You observe a similar effect whenever an ambulance passes by. As the ambulance approaches, the signal is squeezed, resulting in a higher frequency, which you hear as a higher pitch. The opposite happens as the ambulance recedes in the distance: you hear the siren as a lower pitch. Light is a wave and is subject to the same effect. As astronomical bodies recede, the frequency of the light we receive on Earth is lower than that actually emitted by the body. The object appears "redder," so we say the signal is "redshifted" (shifted toward lower frequencies or toward the red). This shift in frequency can be used to infer how fast the galaxy is moving with respect to us. In parallel, we can determine how far away the galaxy is from us based on its brightness and thus establish the Universe's rate of expansion.

Supernovae have been used since the 1960s to establish the rate of change of the expansion of the Universe, but these early measurements were notoriously challenging: separating out the change in the supernova luminosity from that of distant galaxies in the background required the ability to discern subtle differences. However, in 1998 two groups—the Supernova Cosmology Project and the High-Z Supernova Search Team (where "Z" stands for redshift)—simultaneously converged and published a remarkable discovery, which was recognized by the journal *Science* as the "Breakthrough of the Year." To understand the magnitude of this discovery and why it continues to mystify us today, we first need to gain a better sense of what should happen as the Universe expands.

As the Universe expands, several things occur that lead to a decrease in the Universe's energy density (the amount of energy per unit volume). First off, the space between free particles increases, giving each particle more room to itself. The particle density (the number of particles per unit volume) decreases as the volume increases. That is true for all particles, regardless of charge, mass, spin, color, and so on. If a one-meter-cubed box contains ten particles, the particle density is simply ten particles per cubic meter. If that box "expanded" and gradually doubled in size, its volume would be eight meters cubed, resulting in a particle density of 10 particles per 8 m^3, or 1.25 particles per cubic meter.

Now if a particle is massive (that is, it has mass), the energy it carries scales like its mass, as you will recall from Einstein's famous formula $E = mc^2$. Because the mass of a particle is intrinsic to that particle and remains constant as the Universe expands, the energy density of massive particles (their energy per unit volume) will also decrease as the Universe expands and the volume increases. But if a particle has no mass, like the photon, then its energy is determined by its wavelength rather than its mass: the shorter the wavelength, the greater the energy of the massless particle.[1] As the Universe expands, space stretches, increasing with it any notion of distance, including the notion of wavelength, which is the distance taken by a wave to make

1. Einstein's formula $E = mc^2$ is only valid for a particle at rest. However, particles like photons are massless and thus never at rest; in fact, they are always moving at the speed of light. In that case, another relation must be invoked, one which even predates Einstein's formula and was proposed as early as 1900 by German physicist Max Planck. Planck's formula states that if we consider an electromagnetic wave of wavelength λ, the energy contained in that wave is inversely proportional to that wavelength, going as $E = ch/\lambda$, where c is the speed of light and h is the constant of proportionality known as Planck's constant, the same constant we encountered in chapter 4 when discussing the Planck energy scale.

an oscillation. And as the wavelength increases, the energy carried by any electromagnetic wave decreases.

Let's return to the example of our box containing some particles. As the box doubles in size, the density of particles decreases by a factor of eight. If the particles are massive, the energy carried by each particle remains the same as the box expands, and the energy density contained in that box therefore also decreases by this same factor of eight. However, if the particles are massless, the energy carried by each particle decreases by half as the box doubles in size, meaning that the energy density of massless particles decreases by a factor of sixteen. What this example shows is that the energy density carried by light or "radiation" dilutes more rapidly as the Universe expands than the energy density carried by massive particles or what we call "matter," whether visible or invisible dark matter.

When the Universe was relatively young, dense, and hot, the energy contained in light and other massless particles (and massive particles traveling close to the speed of light) was the dominant form of energy, causing it to expand rapidly. This is what cosmologists call the "radiation era." Fortunately for us, because radiation dilutes faster than matter, there came a time, when the Universe was about fifty thousand years old, that light and radiation were no longer the dominant form of energy. Instead, most of the energy in the Universe was in the form of matter, what would later become the clusters of galaxies in which we live. This was the beginning of what we call the "matter era."

During this matter era, the Universe continued its expansion at a slightly slower rate. This is when electrons were finally trapped by protons and neutrons' nuclei, and the first atoms were formed, allowing light to propagate freely. Over the next few hundreds of millions of years, the first clusters of galaxies formed, guided by the gravitational pull from dark matter wells, spreading across filament-like structures throughout our Universe.

Since the Big Bang, which blasted space apart, the Universe has maintained its momentum and continued to expand. Initially expanding at an accelerated rate during the period of cosmic inflation, the expansion then gradually slowed down as it progressed through the radiation and matter eras.[2] Over time, we would expect the gravitational pull between clusters of galaxies and any form of matter and mass present in the Universe to eventually win out and further slow the cosmic expansion—possibly to a complete stop, or even to the point where the Universe would collapse back into itself due to the gravitational pull from all its constituents.

To confirm that the expansion is slowing down, we should look at galaxies farther away from us; by observing faraway galaxies, we are seeing what happened earlier in the history of our Universe. If the expansion is slowing down, that means that it was faster in the past, so we should expect faraway galaxies to be moving away from us at a faster rate than nearby galaxies. This is, in fact, precisely what the Supernova Cosmology Project and the High-Z Supernova Search Team set out to demonstrate, and why their results were so significant.

But rather than confirm that the Universe's expansion is currently slowing down, as we would expect, both groups confirmed the exact opposite. The expansion of the Universe is actually accelerating. This discovery earned Saul Perlmutter, Brian P. Schmidt, and Adam G. Riess the 2011 Nobel Prize in Physics. Today, we have a multitude of additional observational probes at our disposal, all of which confirm that the Universe's

2. There could be alternatives to cosmic inflation, and fascinating models have been proposed, but they all agree on the radiation and matter eras that followed.

expansion is indeed accelerating (though the precise rate of acceleration measured by different probes suffers from tensions). This discovery is vexing and perplexing on many levels, but it is only when we delve deeper into the interface between gravity and particle physics that we will be able to understand why such an observation goes against anything we would have anticipated.

Dark Energy

Throughout the cosmic times, one would have naively expected that the cosmic expansion would gradually slow down and the overall energy density of any element would decrease. This should be the case if the Universe is filled with radiation (light) or matter of any kind, regardless of its form; it could be planets, comets, stars, gas, dark matter, black holes, or neutrinos—all of these elements contribute to the deceleration of the cosmic expansion. To explain the accelerated cosmic expansion, something quite different must happen. The Universe must be filled with a new type of energy that doesn't dilute as the Universe expands (or only dilutes very weakly). We haven't seen or interacted with such energy (yet), so it must be "dark" to us, but it is nothing like dark matter. This quintessential energy has been dubbed *dark energy*.

Since 1998, cosmologists have made remarkable progress in determining what dark energy is not. But the questions of what dark energy really is and what drives the accelerated expansion of the Universe remain unanswered. Dark energy is often described as a phenomenon of "antigravity," which reverses the gravitational attraction between local masses. To describe dark energy within the framework of Newtonian gravity, we would need to fill the Universe with a constant density fluid of negative mass! This doesn't seem promising. In general relativity, things are not so dramatic: dark energy can be described as a fluid with positive energy density, but one with negative pressure. It is the negative pressure that creates the effect of antigravity, forcing galaxies to accelerate away from each other. More precisely, the galaxies continue largely on their own straight-line paths through a curved geometry, and it is the negative pressure that causes spacetime itself to dilute and expand at an exponential rate, carrying the galaxies along with it.[3]

One initially bizarre upshot of this analysis: if we had a box full of dark energy and doubled its size, the total energy contained in the box would increase. But where does the extra energy come from? And doesn't the principle of conservation—the idea that energy can only be transferred into another form of energy, such as heat, but cannot be gained nor lost—rule out this possibility? How, then, can we explain the fact that the total energy of the Universe is increasing over time?

In physics, energy conservation is related to the principle of "time-translation invariance." If the laws governing the dynamics of a physical system do not change with time ("invariant

3. If you are curious, we can mimic dark energy already in Newtonian gravity (although of course it has a much more profound derivation in general relativity). We typically think of the Newtonian force acting on an object of mass m as deriving from the Newtonian potential V, as $F(r) = -mV'(r)$. For an object of mass M localized at $r = 0$, the potential is given by $V = -GM/r$, so that $V' = GM/r^2$, and we recover the gravitational Newtonian inverse square law $F = -GmM/r^2$. If, instead of a localized mass M at $r = 0$, we had a fluid of constant density ρ, the mass contained in a volume of radius r would be $M(r) = 4\pi\rho r^3/3$ and the derivative of the Newtonian potential would be given by $V' = GM(r)/r^2 = 4\pi G\rho r/3$, resulting in an attractive force that increases with distance, $F(r) = -mV'(r) = -4\pi \, mG\rho r/3$. To mimic the effect of dark energy, we need to turn this into a repulsive force, which causes distant galaxies to accelerate away from each other. In Newtonian gravity, this requires an unphysical negative mass density $\rho < 0$. In general relativity, this antigravity property can be achieved with a fluid with positive energy density but sufficiently negative pressure.

under time translations"), then energy will be conserved. This holds true in Newton's mechanics, in Maxwell's theory of electromagnetism, and in special relativity where the geometry of spacetime is flat and unchanged in time. But while these are good approximations here on Earth, our Universe is nothing like that. It is curved, expanding, and evolving in more ways than we can comprehend. Physical quantities do not have to be constant in a time-dependent system, and the energy of the Universe is no exception.

The details underlying the beginning of our Universe still elude us, but most cosmologists would reckon that when the Universe was born, dark energy was so negligible that it accounted for only about 10^{-124} of its total energy. As a percentage, this is about 0.0000000000000000000000000000000000000000000000000000 0000000000000000000000000000000000000000000000000000000 00000000000000000000000000000001% (at this point, you can give or take a few zeros, not many would even notice). This is such a minuscule contribution that it is difficult to comprehend. For comparison, the mass of a single atom of hydrogen is about 10^{-51} that of our entire planet. So, while the impact of a single atom on the motion of our planet is obviously imperceptible and insignificant, it is still incomparably greater than the impact of dark energy at the beginning of our Universe.

The real strength of dark energy is not in its local impact in any one single region of space and time but its constant nature throughout space and time—its cohesion and persistence. Dark energy doesn't diminish during the cosmic expansion and doesn't wear off at the edges of galaxies or clusters of galaxies. Even though the Universe likely started out with a minuscule level of dark energy, all the other forms of energy have been progressively washed away during the cosmic expansion. Today our Universe has grown so very old and has been stretched so much that most elements have been almost completely diluted away. Dark energy, on the other hand, simply stayed constant, biding its time, until over billions of years of existence it steadily came to dominate. Once inconsequential, dark energy now accounts for roughly 70% of the total energy present in the Universe. Baryonic matter, which makes up all visible matter (you, me, all interstellar dust, and every planet and star), accounts for only about 5% of the total energy, while dark matter accounts for the remaining 25%. Light or glight radiation, along with radiation from other massless particles, once accounted for most of the total energy, but their current contributions have been far surpassed. Today, the Universe appears to bend to the rules of dark energy, following its instructions about how to evolve and accelerate.

If we take a moment to reflect, you might wonder if this isn't all too good to be true. Are we really advocating that there's a lot more energy out there that we originally thought? If so, why aren't we harvesting it? And, aside from the fact that we wouldn't be here without it, why hasn't it had a more significant impact on our lives?

Dark energy exists everywhere in the Universe, even on Earth, even behind our ears, and some amount of it fills our lungs every time we breathe. There are, however, a couple of obstacles that prevent us from using it as an ideal source of green energy. For one thing, the only way we can interact with dark energy is through gravity, which is the weakest of all forces. But the main problem is that the density of dark energy is still very small *locally*, meaning at any one point. If we were miraculously capable of harvesting all of the dark energy present within ten meters above the Earth's surface, this would provide enough energy to boil a couple of kettles of water. Aside from a lovely cup of tea, we wouldn't have gained all that much. Even collecting all the dark energy present in a volume encompassing the Earth and the Moon would only provide us with enough energy to launch a few spacecrafts into orbit. Sadly, the market for dark energy farming is unlikely to pick up anytime soon; perhaps that's for the best because dark energy thrives best when left undisturbed.

Even today, the presence of matter completely overshadows the role of dark energy in any "inhabitable" region of our Universe, where galaxies have formed. All the stars, gas, black holes, and dark matter present in the Milky Way are far more important than the total amount of dark energy it contains. This has always been the case and will continue to be so until the end of time. Even when the Milky Way collides with Andromeda, and its graceful spiral is disturbed, dark energy will remain hidden in the background, unheard, unseen, and completely irrelevant to the process.

To appreciate the significance of dark energy, we must leave the habitable regions we are comfortably settled in and travel to the vast empty cosmic voids that separate clusters of galaxies. Without dark energy, these vast regions of space that make up the majority of our Universe would be completely devoid of energy. While matter is primarily concentrated around clusters of galaxies, dark energy is ubiquitous throughout the Universe. This unwavering determination to fill every pocket of spacetime is precisely what allows dark energy to ultimately have a greater impact than any other component of the Universe.

Einstein's Cosmological Constant

When Einstein derived his theory of general relativity in 1915, the expansion of the Universe had not yet been established. Like most scientists at the time, he believed that the Universe was in a relatively stable static state. The gravitational attraction of matter would then necessarily disturb this balance in ways that Einstein and his contemporaries found unacceptable. Luckily, Einstein proposed a quick fix to this problem in 1917, although one he later admitted may have been "the greatest blunder" of his life. The essential idea may sound familiar, reflecting what we instinctively do as scientists when our models fail to meet our expectations: add a new ingredient. This time, the fix did not take the spherical shape of a planet named Vulcan, or planet 9, nor did it take the form of dark matter; but rather it is represented by a Greek letter Λ, also known as the *cosmological constant*. This constant is the only modification to the original equations of general relativity that did not undermine the philosophical pillars on which his theory of gravitation was built—namely, the equivalence principle—so Einstein was perhaps not too unhappy to introduce it.

In the absence of gravity, the cosmological constant merely indicates a reference point. Imagine you want to measure the altitude of a mountain. How should we define altitude? Do you use sea level as your reference point? Or the center of the Earth? There is no unique choice. Although choosing a different reference point changes the numerical value for the altitude of the mountain, the mountain itself, as well as its relation to other objects, is unaffected by the choice. Changing the ground zero reference point causes an overall constant shift in the altitude, but relative altitudes remain the same. The cosmological constant is the equivalent reference point or ground zero for the energy density in our Universe. Nothing is affected by this shift.

Nothing? Well, almost. The gift and curse of gravity, which follows from the equivalence principle, is that *everything* is affected by it, and it affects *everything*. When I said that nothing is affected by this cosmological constant shift, I meant nothing except gravity, which is precisely what Einstein was banking on.

The cosmological constant was meant to compensate for the gravitational effect of matter on our Universe in an attempt to keep it static—which is what he expected it to be.

After the initial eureka moment, however, the scientific community eventually realized that this balancing act wasn't stable. The distribution of matter would have to be completely homogeneous, and the cosmological constant would have to be precisely tuned to compensate for the gravitational effect of matter everywhere. If Λ is even a tad too small, the Universe collapses in a cataclysmic way. But if Λ is just a pinch larger, the Universe speeds up in an unstoppable accelerated expansion. There was no safe, static middle ground, so the cosmological constant appeared to be a massive failure.

The twist, of course, is that, as we now know, the expansion of the Universe is actually accelerating! Far from being Einstein's greatest blunder, this cosmological constant could simply be the missing dark energy, yet another stroke of genius ahead its time, explaining the late time acceleration of the Universe before it was even observed. The value of this cosmological constant has now been determined empirically to excellent precision, even while we remain entirely in the dark about where it comes from. What determines the value of this constant? What is the physical phenomenon that explains its presence?

Diving into the Vacuum

After diving headfirst into a black hole, reliving the moments following our Universe's birth, and contemplating clusters of galaxies flying apart like the letters on a child's balloon, there's one more stop on our guided tour of the Universe. Let us pay a brief visit to the deepest and darkest abysses of our Universe: the cosmic voids denuded of any kind of matter and conventional energy. You might think of these voids, which can stretch for millions of light years, as rather lonely and boring places, but these desolate spaces might just provide the most exciting clue to understanding dark energy on a more fundamental level.

Imagine yourself plunged into the middle of an intergalactic cosmic void, more than a hundred billion billion kilometers away from the closest galaxy, so far from any form of life that light from the closest star would take tens of millions of years to reach you. Imagine how empty you would feel—no human conversation, no gardens to tend to, no pastries, no tea, no seaside benches with a lovely view. To make matters worse, you wouldn't even have an exciting spacetime curvature to appreciate. There, in the middle of this cosmic void, the curvature of spacetime has fallen to the lowest possible level the Universe has ever experienced, the lowest it has ever been since the beginning of time—sixty orders of magnitude below the Planck scale, some fifteen orders of magnitude lower than it is in the empty space of our solar system.

Although we can't actually travel to such a cosmological tundra, perhaps we can re-create the experience here at home by making a huge vacuum chamber. Take a room, or perhaps a closet in your house, and empty it completely. Seal it tightly and remove all the air with the help of the strongest pump in the world. Make sure that you've removed every last random electron, photon, and other fundamental particle. OK, done? Now isolate the chamber within kilometer-wide reflective, concrete walls, ensuring that no light, neutrinos, cosmic rays, or other particles manage to make their way down into your secret chamber. As you congratulate yourself on having created the best vacuum ever, infinitely better than any vacuum encountered in outer space, know that even your prize-winning vacuum chamber would not be entirely empty. It will be filled, not with matter or physical particles, but with energy, the energy of a quantum sea of particles and antiparticles that are continuously being created and annihilated—what we refer to as "vacuum energy."

The idea that something could be created or destroyed in such a vacuum may seem uncomfortable because it goes against the natural intuition that we shouldn't be able to get something out of nothing. But quantum physics allows this to occur in sufficiently small regions of space and for short enough periods of time. The electron, for example, is a real charged particle that can easily be removed from the room by attracting it with an oppositely charged source. Moreover, the electron has an antiparticle called the positron with identical properties, except for having the opposite electrical charge. Quantum physics tells us that there is some probability that a pair of an electron and a positron, which has zero net electrical charge, can be created in the vacuum, at any place and at any time, provided they disappear by annihilating each other (or by annihilating other pairs) in a very brief time determined by Werner Heisenberg's uncertainty principle.[4] We may not be able to directly catch such particles with our eyes or instruments, which is why we are unable to remove them from our vacuum chamber and why we refer to them as "virtual particles," but the effects of the spontaneous creation and disappearance of pairs of these virtual particles have been measured and tested with incredible precision via various experiments.

Some of the most well-known of these experiments take place under the city of Geneva, some 50 km west of where I was born

---

4. The energy in an electron–positron pair is at least $E = 2mc^2$ where m is the electron mass. According to one version of the Heisenberg uncertainty principle, an uncertainty in energy of ΔE implies an uncertainty in time of $\Delta t = h/4\pi\Delta E$, where h is the same Planck constant we encountered earlier. This implies that a virtual electron–positron pair may exist for a time of order $\sim h/8\pi mc^2$ without being detectable.

(Lausanne), in the world's most powerful particle accelerator, the Large Hadron Collider (LHC). At the LHC, protons are smashed together at 99.9999991% of the speed of light when they reach a combined energy on the order of a dozen trillion electronvolts, which is millions of times more energetic than an electron at rest.[5] Correctly interpreting the results of the collisions that occur at the LHC requires understanding how the spontaneous creation of virtual pairs of particles affects these processes.

For instance, without accounting for these effects, we would not have been able to explain how the Higgs boson decays into photons, which is one of the ways the Higgs was detected at the LHC in 2012. The Higgs particle was postulated in 1964 by multiple groups in parallel (including Brout, Englert, Higgs, Guralnik, Hagen, and Kibble) to explain how particles like the W and the Z bosons acquire an (inertial) mass. Think of the Higgs as a sea within which other particles are submerged; the discovery of the Higgs particle is proof that the vacuum is not empty but is filled with at the very least the energy of the Higgs field. This is part of what we call the vacuum energy, or the energy contained in the "emptiest" of spaces, such as our vacuum chamber or the intergalactic void we visited. The Higgs contribution to this nonvanishing vacuum affects every other massive particle, slowing them down and effectively leading to the origin of their inertial mass, but it can also have other effects as we shall soon see.

5. In particle physics, we commonly use the unit of electronvolt (eV) to designate a unit of energy. One eV corresponds to the amount of kinetic energy that an electron originally at rest gains when it is accelerated through an electric potential difference of one volt. One eV corresponds to about 10^{-26} kWh. Bearing in mind that a normal household consumes about 10 kWh per day, the amount of energy contained in 1 eV is beyond ridiculously small from a human perspective. When it comes to the Higgs boson, we will often designate its mass in billions of electronvolts corresponding to a giga-electronvolt, or GeV for short.

Since the Higgs boson is a neutral particle (it has no electric charge), it does not directly interact with photons. However, the Higgs can produce a virtual pair of charged W^+ and W^- bosons, which then interact with photons. This means that the Higgs decays into photons through the intermediary of a virtual pair of W bosons, which is precisely what was observed at the LHC. In fact, the effects of those virtual particles, commonly known as *loop corrections*, are so well understood and are in such impeccable agreement with data at such a high level of precision that not only do we routinely account for the effects of the creation of virtual pairs, we even account for the creation of a virtual pair within a virtual pair within a virtual pair within a virtual pair within a virtual pair. (If you've seen the sci-fi film *Inception*, this would be equivalent to gaining complete control over five levels of a dream.)

This may seem like the kind of thing that is only possible in Hollywood's imagination, but it is the everyday reality for particle physicists. In the particle world, the amount of evidence pointing back to the effect of virtual particles is simply overwhelming; dismissing their existence would require dismissing more than a half-century of scientific progress and invalidating more than a quarter of all Nobel Prizes awarded over the past twenty years. Even though these virtual particles cannot be directly detected, there is no doubt that they continuously affect the scattering of our physical particles and are an intrinsic part of our Universe's reality.

Unlike dark matter and dark energy, virtual particles are not hypothetical new forms of matter or energy. They are simply the omnipresent soul of all known particles, ever-present due to quantum fluctuations. Every fundamental particle of which we are aware—electrons, photons, quarks, neutrinos, W and Z bosons, and so on—can manifest themselves at the virtual level and spontaneously affect physical processes. In some cases, virtual particles can open up new channels that would have been impossible without them, as is the case for the decay of the Higgs boson into two photons, which would be impossible without a virtual pair of charged W bosons. In other cases, loops of virtual particles can affect ever so slightly the probability of a given outcome. Even the Higgs particle, discovered at the LHC in 2012, provides proof that the vacuum is not empty. Rather, it is filled with at the very least the vacuum energy of the Higgs field, and it is precisely this nonvanishing vacuum value that affects every other massive particle, slowing them down and effectively leading to the origin of their inertial mass.

This sea of quantum particles exists everywhere, at all times, in the voids, throughout the galaxies, and even near black holes. This means that the vacuum energy of particle physics fills every innermost layer of the Universe; in its persistence through space and time, it acts exactly like a cosmological constant. In fact, from the point of view of Einstein's classical equations of general relativity, this vacuum energy is indistinguishable from a cosmological constant. So this Λ, originally introduced by Einstein as a quick fix, could simply be the expected vacuum energy from every particle and is thus the most natural and obvious driver of the Universe's accelerated expansion. At the very least this is one of the most widely accepted explanations for what our Universe, and indeed the entire structure of reality, has to offer because it combines the two pillars of modern science in perfect symbiosis. On one side, quantum particle physics predicts the existence of a constant and uniform sea of vacuum energy. On the other side, the equivalence principle leads to the laws of general relativity, which predict that the Universe should experience an accelerated expansion in response to this vacuum energy. Together, they depict a coherent and consistent picture of the Universe's history, and one that accords with the initially perplexing observation that it is expanding at an accelerating rate. At first sight, everything seems to be working perfectly. Nature, unfortunately, is never quite so straightforward.

The Greatest Discrepancy in Scientific History

When I find myself in the middle of a set of intricate calculations, which have taken months to complete, and things appear to be working remarkably well, fitting my expectations and heading in the right direction, I know it's often best to stop and wait until the next day before carrying out the final steps. At least, that's what I have come to learn will result in the most pleasant evening for myself and those around me, leaving me blissfully unaware of the complications and issues that await me once I perform my final calculations. And that's just how we left things in the previous section, where it seemed almost certain that the vacuum energy of particle physics is naturally present everywhere in the Universe, serving as a natural cosmological constant and explaining the accelerated expansion of the Universe. All that remains for us to do is to complete our calculation. To convince ourselves (and the rest of the scientific community) that the accelerated expansion of the Universe is caused by the vacuum energy of particles, we simply need to compare the amount of energy density we expect the vacuum to carry with the amount of energy density required to explain the accelerated expansion of the Universe. With the holy grail so close at hand, I'm sure a little math and a few numbers scattered here and there won't cause you to turn back now.

As we've seen, the actual amount of dark energy at any point in the Universe is rather low, about 10^{-33} kg per cm^3, or 10^{-12} eV^4

in particle physics units—barely enough to brew a few cups of tea if we were able to harvest all the dark energy around the Earth's surface. That's the amount of dark energy density that we require to explain our observations about the accelerated expansion of the Universe.

Now let's see how much vacuum energy particles can provide. Consider a particle of mass m. It carries energy E that scales with its mass (recall Einstein's famous law $E = mc^2$). Every particle has its own antiparticle, which has the same identical mass m. Heisenberg's uncertainty principle allows for a particle–antiparticle pair to pop out of the vacuum for a time of no more than $\Delta t \sim h/E$, where h is Planck's constant.[6] This time Δt scales as the inverse of the energy of the particles and hence as this inverse of their mass. In this time Δt, these particles can at most move a distance $d = hc/E = h/cm$ and so at most occupy a volume that scales as the inverse of the mass cubed. If that went over your head, take my word that this means that, in the end, for every particle of mass m, the vacuum should be filled by a sea of virtual particle–antiparticle pairs whose energy density scales like the mass to the fourth power.[7]

Take the electron, for example. Sir Joseph John Thomson discovered this particle in 1897 after decades of speculation. A single electron has a very small mass by human standards, about 10^{-30} kg, but for the vacuum this corresponds to an energy

6. To be precise, the uncertainty in time Δt associated with producing a particle–antiparticle pair at rest of mass m—that is, with total energy $E = 2mc^2$ is $\Delta t = h/(8\pi mc^2)$.

7. The energy density is an energy per unit volume. In particle units, we have seen that everything can be related back to energy or mass (those are related by $E = mc^2$). To probe small distances, we need higher energy so energy scales as 1/length and 1/volume scales as energy-cubed. This means that the energy density scales as energy to the fourth power, or equivalently mass to the fourth power, $E/d^3 \sim c^5 m^4/h^3$.

density of 10^{-30} kg ~ 0.5 Mega electronvolts to the fourth power, or about $(0.5 \times 10^6 \text{ eV})^4$ ~ 10^{22} eV^4. A ginormous amount! Following a similar comparison to that made by British physicist Sir Oliver Lodge in 1907, this would be as if "the total output of a million-kilowatt power station for thirty million years, exists permanently, and at present inaccessibly, in every cubic millimetre of space" [14]. It is *inaccessible*—unless we count its effect on gravity. In terms of its impact on the accelerated expansion of the Universe, it would be more than thirty-four orders of magnitude larger than the amount of dark energy required to explain current observations. In other words, introducing this electron vacuum contribution as a source of energy into Einstein's equations would result in a curvature that is so large it would not be viable.

Although the modern connection between the cosmological constant and the quantum vacuum energy was made by Soviet physicist Yakov Zel'dovich in 1968, the idea that the vacuum carries energy with cosmological implications can be traced back to arguments put forward by German chemist and physicist Walther Nernst in the 1920s. Remarkably, in response to this suggestion, German physicist Wilhelm Lenz estimated in 1926 that the resulting curvature of spacetime induced by vacuum energy of waves with frequencies as high as that of the electron mass would be so large that the resulting Hubble or observable radius of our Universe (the distance beyond which we wouldn't be able to see beyond the horizon) "would not even reach to the moon" [15, 16]. Wolfgang Ernst Pauli reached the same conclusion in 1933 [17, 18].

Today, even more massive particles are known to exist. If we take into account the Higgs boson's contribution to the vacuum energy, for example, the resulting curvature of the Universe would be so large that the observable radius of our Universe would barely reach the centimeter threshold. Taken at face value, this would mean that we wouldn't be able to see anything beyond this distance, and that anything that isn't bounded to us via electroweak or strong forces would literally fly away from us faster than we can see. The fact that you can read this book without it escaping at a speed faster than light indicates that this is not the reality in which we live. The vacuum energy that we have estimated cannot be the dark energy that accelerates the expansion of our Universe.

But where did we go wrong? Here we have two very well-tested and trusted theories. On the one hand, we have quantum physics, which governs the realm of particle physics with its sea of virtual particles whose contributions have been established with impeccable precision in particle physics experiments; on the other, we have general relativity, whose predictions have so far never betrayed us, even after more than a century of scrutiny. Yes, general relativity told us it will break down at some point, before reaching black hole singularities or the Big Bang singularity, but not on cosmological scales. Could it break there as well?

The real question we face today, what we call the "cosmological constant problem," is not about the nature of dark energy that causes the accelerated expansion of the Universe, but rather it is about why it is accelerating so slowly. Why is the observable Universe allowed to be so large when the vacuum energy should have curved it fiercely? In other words, why does dark energy *only* comprise 70% of the total energy when we would have expected it to make up so much more? Or, more jarringly, how is it the case that "ordinary" matter (that's you and me) is able to account for as much as 5% of the total energy currently present? Of course, we wouldn't be here to ask ourselves that question if it hadn't been the case, so perhaps nature is just so kind as to provide us with the proper environment for us to survive. Such an anthropic answer is always possible, but it does not explain how our Universe came to be the way it is.

Another possibility is that when applied to the context of the vacuum, our expectations from particle physics are simply incorrect. Perhaps there is no such thing as quantum vacuum energy. This was close to the line of thought that followed Lenz and then Pauli's realizations nearly a century ago. Even though Pauli was one of the founders of quantum mechanics, he believed that the idea of quantum vacuum energy should be rejected.[8] The difference between then and now, of course, is that we have since observed that the expansion of the Universe is, in fact, accelerating. Something must be pushing it to do so—and if it is not vacuum energy, then what else? Setting the vacuum energy to zero or simply ignoring it and postulating the presence of a new dark energy source for the accelerating of the Universe is another way forward, but this does not address the core of the problem: why isn't our Universe curving strenuously under the influence of vacuum energy?

The Universe communicates with us in a variety of subtle ways. Are the mysteries we are currently facing a clue to a deeper truth? Looking for signs of new physics has been at the heart of every breakthrough, but this simple and innocent question is currently causing consternation within our scientific community. In this book so far, I have told the story of gravity with the privilege of hindsight, drawing on and recasting decades of research. We are, however, about to journey off the map. Chapters 6 and 7 chart an adventure that is in the process of unfolding, a quest that we have the luxury to steer ourselves.

8. Pauli refers to the quantum vacuum energy as the "zero-point" energy, which is still what we sometimes call it today. In the absence of gravity, the vacuum energy is a zero-point energy because it has no or "zero" point of contact with anything else—that is, we do not interact with it. However, we feel its gravitational effect, and when including gravity the vacuum energy is no longer a zero-point effect.

CHAPTER 6

The Graviton, What a Particle!?

In 2021 the European Space Agency (ESA) announced its fourth astronaut selection. It wasn't long before people asked me whether I would jump at the opportunity to try again. If not, why didn't I try my prospects in North America or within the private sector? Better yet, why not take matters into my own hands and develop the space sector in other ways? Why was I letting this one obstacle derail what had been my lifelong dream?

It certainly wasn't laziness, nor lack of commitment. It wasn't even lack of imagination. No, the answer can be found in the beauty of falling. The key is not only the amount of courage and determination required to leap into the void, nor how graceful the jump is, nor how stunning the view from above. The key to falling, through life and through the sky, is to enjoy it for what it is, savoring the sensation of weightlessness while knowing when and how to stop—to pull the ripcord, release the parachute, and channel the rush of free fall into another direction.

In the wake of my latent tuberculosis diagnosis, it was time for me to seize the opportunity for a new adventure, one that would continue along a different path—a path that was as uncertain as it was challenging, just like the one that brought me within a stone's throw of outer space, but one that offered the chance to contribute to the discovery of something new, making the journey worth it all.

So Far, So Good

General relativity has successfully run its course over dozens of orders of magnitude, revealing beauty in the structure of space-time and enlightening, or perhaps englightening, the very foundations of nature. Just as during the joy of free falling, our exploration of the Universe with general relativity may feel so perfect and serene that we could easily lose ourselves, forgetting what is to come. However, in predicting its own failure, general relativity has always made it clear that this blissful ride would not last forever. What comes next is bound to be as unpredictable as it is challenging.

Fortunately for us, our ride is ongoing. General relativity has been tested to incredible precision within our solar system and has proven to be an unfailing guide within our immediate environs. But what happens as we leave the comfort of our home star, our galaxy, or even our local galaxy cluster? Can we trust general relativity to guide us as we travel to the limits of our Universe? The majority of the astrophysical and cosmological community would say yes! *So far, so good.*

In fact, over the past twenty years, our current cosmological paradigm has allowed us to achieve exceptional progress on so many questions about our Universe. We have been able to successfully explain the evolution of our Universe from a fraction of a second after the Big Bang to its subsequent evolution over billions of years. We have been able to predict the pattern of light shining from the afterglow of the Big Bang. We have been able to describe the formation and distribution of clusters of galaxies, and the abundance of elements in the Universe, all in complete agreement with current observations. However, despite these many successes, the cosmological paradigm does demand of us a few tiny sacrifices, a dogma we must accept to make further progress. In the making of our cosmological paradigm, we must take special care to include certain ingredients and follow a particular recipe:

1. With a thin brush, start by clearing out the cosmological constant problem and assume that one can simply ignore the gravitational implications of the quantum vacuum energy of known particles, which would otherwise cause the Universe to curl to less than a centimeter in size.

2. Next, fill every pocket of the Universe with a dark type of energy, a fluid that has physical properties unlike anything we have ever measured and accounts for 70% of the total energy in the Universe.

3. Then further accept that 95% of our Universe is invisible to us, that most matter is living a secret life parallel to us.

4. And for the last step, make sure to treat as insignificant the discrepancies between various measurements of the expansion rate of the Universe—measurements that we have every reason to believe are accurate—and hope that they will eventually simply disappear, even though the current odds that this tension is just an error are less than one in three million.

If we believe the current cosmological paradigm, in which general relativity is the appropriate description of gravity in the Universe's most isolated environments, even though no measurements nor observations have yet confirmed it, we must accept this procedure and take these ingredients for granted. But what if these assumptions are too much of a stretch? What if . . . and this may sound crazy, but what if there is more to gravity on large cosmological scales than what Einstein's general theory of relativity reveals? Perhaps trusting general relativity in our cosmological abysses is like expecting fish in the deepest reaches of the ocean to look and behave identically to those in our comfortably warm tropical seas. Perhaps our enchanting fall is coming to an end, and it's time to try something new and explore a new path for gravity. If we dared to do so, what might we discover?

Gravity with a Sense of Humor?

For nearly every scientific topic, there are certain aspects on which the scientific community generally agrees or converges, and certain aspects that generate tension and skepticism. So far, the material I've presented in this book falls largely on the "accepted" end of the spectrum. Little of what I've said would be regarded as contentious or controversial. This is not to say that I have presented everything in the traditional light, but I purposefully haven't introduced any views that fall beyond the realm of scientific consensus. This is about to change!

In this chapter, I will share a theory about what could replace the seemingly outlandish assumptions outlined above. This theory is one that I have developed in collaboration with my incredible colleagues, but it is not accepted as definitive by the majority of the scientific community. Even *I* am not entirely convinced we should accept it. Yet it is one that I believe is worth investigating because there is a chance, however small, that we may learn from it and that it may help us develop a better picture of reality. Even if the theory fails, the insights we gain through our investigation will teach us much about how unique general relativity is, about the true origin of cosmic acceleration, and about how to go about testing gravity. Perhaps most importantly, exploring this alternative provides us with the opportunity to ask ourselves new questions and to appreciate the laws of nature from a different perspective.

Einstein's theory of general relativity is built under the premise of the equivalence principle, which we have outlined in earlier chapters. According to the equivalence principle, everything in the Universe should couple equally to gravity and in turn have a similar effect on the structure of spacetime. This principle is at the heart of our cosmological/vacuum energy problem because it demands that the vacuum energy has a gravitational effect and dictates that this effect should completely reshape the structure of our Universe, to the point that it would be unrecognizable to us.

It is no coincidence that the word "gravity" is often associated with the idea of being profoundly serious. At the scales with which we are familiar, gravity is the most serious of phenomena, overpowering everyone and everything with the precise same importance, attention, and dedication. In fact, as described both by Newton and Einstein, gravity is a phenomenon with infinite range, affected by everything with the same serious strength. While the intensity of the gravitational pull decreases as the square of the distance between two objects, there is no distance so great that gravity lets go. Two objects, even if placed at opposite ends of the Universe, would still technically exert a small pull on each other. This implies that the entire Universe should have a gravitational effect on itself, a realization that lies at the heart of the cosmic acceleration problem. How can the Universe's expansion be accelerating at the observed rate when we would expect either the gravitational pull of its own mass to slow it down or the vacuum energy to cause it to accelerate much faster? Without unnatural fine tunings, in which we adjust the vacuum energy to implausible small levels, general relativity is so intransigent that it admits little room for anything between these two outcomes.

But what if, deep down, gravity was not so solemnly serious? Could it have developed a very well-hidden sense of humor over the course of its tens of billions of years of existence? And, after experiencing its own fall, could gravity have a healthy sense of its own limits? What if gravity, appreciating that the time has come, could simply let go? If so, this could explain how a seemingly ginormous vacuum energy could be consistent with the evolution of our Universe. We would not need to disregard the existence of vacuum energy, nor tune it to miraculously and inconceivably small levels. Instead, according to this picture, the effect of vacuum energy on our Universe, as mediated by gravity, is simply relaxed over space and over time.

If all of this sounds implausible, or even a bit like gibberish, rest assured that you are not alone. Many scientists have previously considered the possibility that gravity could "let it go" (more precisely, that it might have a finite range and not interact with everything in the same stern way), but they ultimately rejected the idea because making sense of it proved so difficult. To understand how this idea would work, as well as the problems that previous scientists have encountered, we must return to gravity's hidden force.

We have seen that gravity, while corresponding to the manifestation of the curved geometry of the spacetime in which we live, nevertheless carries a genuine force, manifested in the tidal effects that can be as subtle and gentle as those now routinely observed in gravitational wave observatories, or as dramatic and fatal as those that would tear us apart if we fell into a black hole. This force is communicated via glight or gravitational waves, just as the electromagnetic force is carried by light or electromagnetic waves. We now know, thanks to the insight of many brilliant scientists such as Stanley Deser, Richard Feynman, and Steven Weinberg, that general relativity is the unique, consistent way to describe a massless glight wave. In other words, simply demanding that gravitational waves are infinitely serious leads us unambiguously to general relativity, without ever mentioning the equivalence principle.

Being as light as glight sounds amazing, if not liberating. I wouldn't mind it myself, especially when I'm rushing to catch my bus. But being too light isn't always that much fun. The absence of mass, or masslessness, implies that gravity must be a stern law, interacting with everyone in a strict manner for eternity, never slowing down and never letting go. The gravity of this phenomenon lies at the core of the cosmological constant problem. But if gravity was allowed to relax, it would not need to take the vacuum energy so seriously. The vacuum energy would only affect the Universe for a finite time and over a finite distance. For this to happen, however, gravity cannot be as light as in general relativity; it cannot have an infinite range. Rather, it has to be *massive*.

The notion that gravity carries a mass or is massive may, at first, appear paradoxical. We typically associate massive objects with the generation of a gravitational field, but we don't think of that gravitational field itself as being massive. Being "massive" in this context simply means that glight, or more precisely the particles that comprise glight, have an inertial mass (as we discussed in the first chapter), not that gravity is gargantuan or that the mass of gravity attracts itself.[1] An element with larger *inertial*

1. Even in general relativity, glight affects the curvature of spacetime around itself (a phenomenon known as backreaction), so in some sense gravity already attracts itself (or self-interacts) in general relativity, although that effect, massive or not, is typically very small. The same phenomenon occurs in massive gravity with a similar overall effect in most situations.

mass is more difficult to move; it is more *inert*. Consequently, a massive graviton (as in massive gravity) is not as eager to move and tires more easily than a massless graviton (as in general relativity). Just as attaching a ball and chain to your leg would impinge your range of travel, attributing inertial mass to gravity would affect how far and for how long the gravitational force would carry its effect. Now we're not talking about something as dramatic as ball and chain for gravity, that would be cruel (let alone unviable for us). What we need is a gentle way to slow it down. If gravity carried a small inertial mass, it would be forced to come to a halt and let go, giving gravity a finite range. Its concealed "sense of humor" would allow it to unwind. As a result, the acceleration of the Universe would be far less dramatic than predicted by general relativity, and our observations could potentially be compatible with the predictions of particle physics.

The idea of a force carrying a mass is, in fact, a common phenomenon in nature. We wouldn't be here without it. Consider, for example, the weak nuclear force, which, as its name implies, is weak in strength. The weak force is essential to life. It is responsible for nuclear fusion, which allows hydrogen to decay radioactively into helium, the Sun's power source. The effects of this weak force are now routinely harvested for our own benefit, such as in medical positron emission tomography (PET) scanning, which carefully monitors the radioactive decay of various isotopes injected into our body and allows for accurate medical imaging.

If you aren't an expert in nuclear medicine imaging or a solar nuclear physicist, you'll be forgiven for feeling a little disconnected from this weak force. It is precisely because this force is *weak* that we aren't as familiar with it as we are to its close cousin electromagnetism. The weakness of this force is directly related to the fact that the particles that mediate it, known as the W and

Z bosons, are quite massive. These particles, along with others, were predicted in the 1960s by Abdus Salam, one of the fathers of the standard model of particle physics as well as the founder of the theoretical physics group at my home institution, Imperial College London. His discoveries, for which he was awarded the Nobel Prize in 1979 with Steven Weinberg and Sheldon Glashow, were instrumental in developing the standard model of particle physics and the understanding that forces can be massive or, in other words, that force-carrying particles can have mass. Incidentally, Salam was also the first scientific Nobel Prize laureate from an Islamic country.

The precise way in which bosons can acquire their mass was discovered by another Imperial College professor, Sir Tom Kibble, together with several other researchers we mentioned previously in the other miracle year, 1964. In the same year, at least six researchers semi-simultaneously recognized that there must be a new particle responsible for giving the mass to the force carriers, predicting what we now know as the Higgs boson (as discussed in chapter 5). Kibble later showed how the masses of bosons were intrinsically linked to the breaking of symmetries and how some bosons such as the photon could remain massless.

To understand the concept of symmetry breaking, imagine you are skating on a perfectly circular frozen lake, surrounded by a forest. As you skate on the lake, you may slide in any direction—north, south, east, or west—as you wish. Because all directions are equivalent, they are "symmetric." If you stand in the middle of the lake, you will be at a point of perfect symmetry: whichever way you look, everything will look exactly the same, with the lake in front of you and the forest beyond. Imagine now that encircling the lake is an iced track, following its circumference but with curved ramping on the sides, the kind of track suitable for a bobsled. Once you are on that track, you can slide freely along the circumference, but the ramping along the sides prevents you from changing course, let alone reaching the point of perfect symmetry in the middle of the lake. The symmetry has now been spontaneously broken: if you look left, you see the lake; if you look right, you see the forest and nothing else.

What does this skating analogy have to do with particle physics? At the beginning of our Universe, all the particles were merged into a very symmetric state. These particles were effectively massless as it cost no energy to move in any direction. As the Universe cooled down, its "icing pattern" evolved, and today it looks much like the track encircling the lake. We are free to move along the direction of the track, as photons do, while moving along the ramp is much harder, costing energy/mass—that's what the W and Z bosons are experiencing.

We now have direct evidence that the Higgs boson is responsible for creating this ramped bobsled-like track, and, in doing so, giving mass to the W and Z bosons. In this process, the Higgs boson has broken the symmetry of the electroweak force, thereby limiting the weak force's range. If the Higgs boson were not present, the W and Z bosons would be massless, like the photon, and the "weak" force would no longer be so weak, as it would have an infinite range. This would have catastrophic implications, undermining all nuclear physics as we know it, affecting hydrogen combustion in the Sun, and compromising the entire solar system in the process. We can be grateful to the Higgs boson for acting as a kind of ramping glue, preventing the W and Z bosons from ever propagating very far or for very long, ensuring the weakness of their force.

If glight had a mass, it would have to be much lighter than that of the W or Z boson, or gravity would already switch off at the nuclear level. However, if gravity is sufficiently light, its range could be long enough to behave essentially the same as in general relativity over the distances with which we are acquainted—whether on Earth, in our solar system, in our galaxy, or even within clusters of galaxies. Everything would look roughly the same at those scales. Only at large distances, over tens of billions of light years, at distances as great as our Universe itself, would the effects of its mass kick in and change the way gravity behaves. The impact of vacuum energy on those large distances and over long periods of time could be sufficiently tamed by the weakening of gravity. This would imply that while the true vacuum energy may be as large as predicted by particle physics, its effect on cosmological scales would be sufficiently relaxed to be consistent with the Universe as we presently observe it.

This tantalizing possibility would allow us to reconcile our expectations from particle physics with gravity, as well as provide a natural explanation for the Universe's accelerated expansion without needing to postulate the existence of a new dark energy filling the Universe. This was, at least, the original idea behind our work on massive gravity, building on the work of many eminent scientists. The modern incarnation of massive gravity emerged from a flurry of new developments in 1999, spurred by the excitement of the discovery of cosmic acceleration. Being massive (at least slightly so) is a positive aspect that allows gravity to finally catch a break and relax after a journey that may have spanned billions of years. You might call it lazy or even a bit tired out, but I'd rather think of it as witty. Gravity has been teasing us in a variety of ways, and while we have been dazzled by its symmetry and inner beauty, perhaps it is time we embraced its sense of humor—the inside joke it has been harboring all along.

The Ghost of Massive Gravity

While the idea of massive gravity has gained new steam in recent years, the concept itself is almost as old as general relativity. The idea that particles closely related to the graviton could carry a mass dates back to Wolfgang Pauli, the same Austrian-Swiss physicist who confirmed in 1933 that, according to general relativity, the expected vacuum energy of the electron would have a dramatic effect on the Universe.

In 1939, Pauli and his former student Markus Fierz were exploring the effect of the spin of a particle. Spin here refers to the internal rotation of an object; for instance, as the Earth orbits the Sun, the Earth also spins on its own axis. This motion is referred to as angular momentum, and spin is the angular momentum that an individual body has even when its center of mass is at rest. Interestingly, many fundamental particles have a tiny quantized amount of spin, and Pauli himself was instrumental in introducing the concept of spin into quantum mechanics.

The notion of spin is sometimes introduced as related to the (quantized) angular momentum of a particle, or how fast it spins about itself. This description is correct for large angular momentum, but what matters for us is that spin is a way to describe how a particle behaves. Just as a particle's inertial mass determines how it reacts when we move it, its spin characterizes how it responds when we rotate it. Take a uniform table tennis ball, for instance. If we spin it on its axis, it will always look the same. No matter which way we turn it, it will look identical. If a particle looks the same under any spacetime rotations, we say that it has "zero spin," or spin-0. Now, instead, consider a perfectly spherical apple with the stem on top. If we flip it over, the stem will be at the bottom, and it won't look quite the same. From its starting position, to see the stem back on top again, we must rotate it a full 360 degrees. In the language of quantum physics, we say it has spin-1.[2] The Higgs boson is an example of a (massive) spin-0 particle, whereas the photon is an example of massless spin-1. The carriers of the weak nuclear force, the W and Z bosons, are examples of massive spin-1 particles.

The graviton is a particle with even more spin. It has spin-2, twice the amount of internal angular momentum as a photon. It looks the same after half a turn or 180 degrees. According to general relativity, the graviton is a massless spin-2 particle. We saw in chapter 3 (see figure 3.3) that massless spin-2 waves, or what we called glight, can come in two different types of polarizations. In fact, polarization simply refers to the various directions in which the spin can point.

Fierz and Pauli were curious about how to describe particles with arbitrary spin. Particles of spin-0, spin-½, and spin-1 were already relatively well understood, so they turned their attention to spin-3/2 and spin-2, considering for each the case with zero and nonzero inertial mass. The spirit of their work was not to challenge general relativity or even to explore alternatives, but simply to generate a consistent classification of all possible types of particles that could be used to describe nature and their properties. They realized that massive spin-2 waves could come in up to six different polarizations, as shown in figure 6.1. The first two polarizations shown in this picture are those we discussed earlier, which are present in general relativity and have been detected by gravitational wave observatories. What Fierz and Pauli could see was that the other four polarizations

---

2. To be more precise, the apple or particle will have helicity-1 along that specific axis of rotation, and it may still have helicity-0 if we flip it over along another axis. However, the ability of the apple or particle to have up to helicity-1 along a direction (and not a higher one) implies that this particle is of spin-1 character.

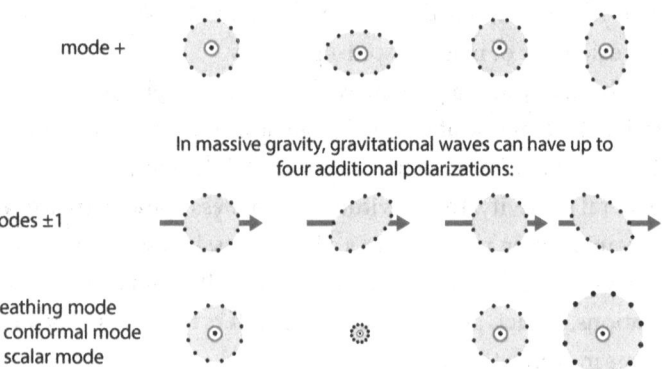

FIGURE 6.1. All possible polarizations carried in principle by massive gravitational waves. Think of the dots as separate beads first placed along a circle. As a gravitational wave passes through, it will distort the spacetime between the beads, giving the impression of a slightly different shape. The arrow represents the direction of propagation of the gravitational wave. The symbol ⊙ indicates that the wave is passing directly through this page. For reference, these waves have been depicted with an unrealistically large amplitude of about 0.5 so that we can get a sense of their effects. *Source:* Claudia de Rham, "Massive Gravity," *Living Reviews in Relativity* 17, no. 1 (2014): 7.

(the two ±1 modes, the breathing mode, and the longitudinal mode) were only possible when the spin-2 particle had a non-vanishing inertial mass. Most of these polarizations require the graviton to speed up or slow down, but since in the massless case (as in general relativity) the graviton is forced to propagate at the speed of light, it does not have the freedom to move along these other four polarizations. However, if gravity departed from general relativity—for instance, if it were massive—glight could come in all sorts of new polarizations.

If nature were an ice cream shop, having more flavors to choose from would mean more satisfaction and enjoyment—unless among the various flavors to choose from, one was so revolting that it had to be avoided at all costs. Then be careful what you wish for! That is exactly what occurs with massive spin-2 flavors. In the ice cream shop of nature, think of the two ± 1 polarizations as new exotic and subtle flavors: they affect distances both along the line of propagation of the wave and along the plane orthogonal to it. Those modes are relatively harmless and difficult to detect. Another mode carried by massive spin-2 waves is the one we encountered earlier in chapter 3, the so-called breathing (or conformal) mode. This new flavor is common in non–general relativity ice cream shops and much harder to miss. If you were to distribute beads along a circle, as represented by the dots in figure 6.1, as this new glight flavor passes through the beads would still form a circle, but the overall enclosed area would oscillate. The breathing mode is still much harder to detect than the standard polarizations, although it gives rise to many intriguing signatures of massive gravity, which we will explore in chapter 7. Before we get into these exciting details, there is one last mode we must mention. In fact, we should be terrified by it! This is the "longitudinal" mode or the *ghost of massive gravity*!

The term "ghost" may appear to have been plucked from a poorly written science fiction novel. That is neither my intention nor my own doing—this is the accepted scientific terminology. In physics, ghosts are particles with negative energy, which are incompatible with the very structure of space and time. The issue is that once we admit ghosts within our picture of reality, a healthy ordinary particle can gain energy by simply "borrowing" the necessary amount from the bottomless supply of negative energy ghosts. Who wouldn't succumb to the temptation? This means that electrons would no longer have to remain peacefully around their atom nuclei; instead, they could borrow energy from ghost particles to become as excited as they desire. Quarks inside the nuclei of every atom would do the same, destroying the structure of matter. Before you know it, every particle in your body, every particle in the solar system, and every fabric of spacetime that connects the Universe would follow their lead, destroying everything we know along the way. I may not have had much of a survival instinct in my younger years, which may have aided many of my more daring endeavors, but the prospect of ghost particles terrifies me.[3]

Ghosts are both intriguing and terrifying, but above all ghosts simply don't exist (or so I tell my daughters and myself when I put them to bed). In their 1939 work, Fierz and Pauli recognized that the equations for massive spinning particles had to be carefully constructed to exorcise the ghosts. If this ghost, or longitudinal mode, existed, it would correspond to the wave slightly speeding up and then slowing down, oscillating faster and slower at the wave's frequency [19]. If glight has no mass, it always travels in the vacuum at exactly 299,792,458 m/s in all frames of reference, so speeding up and slowing down along the patterns of this longitudinal mode is simply impossible. As a result,

---

3. This type of ghost instability is not to be confused with a simple downhill slope instability, also called tachyonic instability. Tachyonic instabilities occur, for instance, in the Higgs potential during a phase transition or what we earlier called the change in "ice structure." In comparison with ghosts, tachyonic instabilities can be relatively harmless and just represent a change of state or vacuum. Ghosts, on the other hand, correspond to the absence of any stable vacuum and would lead to instantaneous decay of all the particles we know.

Einstein unwittingly exorcised the ghost by constructing an infinite range gravitational theory satisfying the requirements of special relativity.

In massive gravity, however, glight no longer needs to travel at the speed of light, so these terrifying longitudinal oscillations *can* occur. As they fluctuate from one speed to another, the flow of time is distorted in a nontrivial way. Because the concept of energy is inextricably linked to the concept of time, if you start messing with time—as this longitudinal wave does when it continuously oscillates, accelerating and decelerating—you also mess with energy, which can then become negative. The more this longitudinal polarization accelerates or decelerates, the more negative the energy becomes. That's when the ghost makes its apparition.

This may all sound quite supernatural, but the mathematical description of these ghosts was well understood more than eight decades ago, when Fierz and Pauli made sure to avoid them in describing how massive spin-2 particles behave on their own. However, Fierz and Pauli were interested in classifying particles, not putting forward a gravitational theory: their spin-2 field was not meant to control the structure of spacetime and dictate how curvature affects anything else trapped in its sway. By attempting to promote these ideas to a full-fledged gravitational theory, we are confronted with quite a challenge. For one thing, in building up this theory, the spin-2 particle is not simply living its own isolated life. Instead, it is gravity's living soul, hidden within the curvature of spacetime and connected with everything else in the Universe. This means that the spin-2 particles must not only be self-consistent on their own, they have to remain healthy no matter how much other particles jiggle them and no matter how curled and twisted spacetime becomes, ensuring the absence of ghost particles at every step of its construction. And this is where the idea of massive gravity ran aground, as scientists deemed it an impossible task. It's not that scientists didn't try to construct such a theory; on the contrary, it's just that every attempt seemed to fail. Disheartened by this failure, in the 1970s some of the fathers of modern gravity proved a series of so-called no-go theorems purporting to establish with absolute certainty that massive gravity simply could not exist without awakening the dramatic ghost. And this impossibility was proven again and again until the mid-2000s.

By the time I started my PhD, this result had become so ingrained in our understanding of gravity that I would never have thought of questioning it. It was not my intention to overturn such well-founded results. After all, as a woman in physics who couldn't possibly know the answer to any reasonable question, I knew better than to attempt the impossible and hasten my expulsion from the theoretical physics community. Rather, my journey to this iconoclastic idea began by working on the idea of extra dimensions of space, sometimes quite large extra dimensions.

From Extra Dimensions to Massive Gravitons

If you are standing on a station platform waiting for your train, you know that it can only come from the left or the right. There is no other choice. Trains are confined to their tracks and live in a one-dimensional world. However, if you're on a boat and know how to harness the winds, there is much more freedom. You can go forward or backward, left or right; boats live in a two-dimensional world. Flying opens yet another dimension: altitude. Planes live in a three-dimensional space. If there is one thing that I would love to do even more than venturing into space it would be traveling along another dimension. My first choice, of course, would be to travel back and forth in time, the fourth dimension. Sadly, that's also quite impossible. So let me be reasonable for a change and simply ask to travel in an extra dimension of space.

The possibility that there could be additional dimensions of space is not entirely confined to the realm of science fiction. The idea first emerged at the beginning of the twentieth century and resurfaced in the late 1980s, the 1990s, and throughout the 2000s, motivated by some breakthrough developments in string theory and supergravity theories. These supersymmetric theories were developed during the search for a "grand unified theory," or what we would now call a high-energy completion of gravity, valid beyond the Planck energy scale. The most ambitious approach, string theory, attempted to explain gravitons as quantum oscillations on fundamental strings from which space and time are made. In effect, the strings of string theory are the very thread from which the fabric of space and time are sewn. Following the pioneering work of Michael Green and John Schwarz in the 1980s, several types of string theories were initially developed, each potentially describing gravity at the quantum level. At low energies, each string theory looked indistinguishable from a special extension of general relativity called supergravity, in which the massless spin-2 graviton of general relativity was supplemented by a fermionic partner of spin-3/2 called the gravitino. Interestingly, they all shared an unusual feature: to be consistent, these various string theories all required the existence of additional spatial dimensions, beyond the three we experience every day. There appeared to be no choice in this matter!

The concept of extra-spatial dimensions was first proposed by Theodor Kaluza in 1919, only four years after Einstein announced the final version of general relativity, in an ambitious and rather beautiful attempt to unify the forces of gravity and electromagnetism. Following later work by Oskar Klein, those additional dimensions are typically thought to be wrapped up into small compact bundles, so small that we are unable to appreciate their extent. String theorists and supergravity theorists of the 1980s sought to rely on a similar Kaluza–Klein mechanism to explain why we hadn't yet seen the missing extra dimensions. This remained the standard picture for roughly fifteen years.

When I was starting my PhD, a new possibility emerged, one that was even more exciting and testable: in addition to these small, compact, extra dimensions, wrapped up in the manner proposed by Kaluza and Klein, there could be at least one "large" extra dimension. This idea was inspired by recent proposals for a new unified theory known as M-theory, proposed by Edward Witten, which could potentially unify and reconcile the various types of string theories discovered thus far. (The label "M" is a placeholder, which could purportedly stand for "mother," "magic," "mystery," "membrane," or perhaps something entirely different.) While superstring theory required nine dimensions of space (ten of spacetime), M-theory proposed that spacetime was eleven-dimensional, the maximum number of dimensions allowed by supergravity theories. The possibility of at least one large extra dimension piqued the interest of many physicists.

These ideas were so influential that they sparked most of the theoretical developments that are still being explored today, such as the powerful notion of holography, developed by Juan Maldacena, which connects gravitational theories to nongravitational theories that live in lower dimensions. Large extra dimensions were also proposed as a potential solution to the so-called hierarchy problem, or why gravity appears to be so much weaker than the other forces. In these models, our entire Universe would be confined to a "brane" (membrane or surface) embedded in the large extra dimension. Everything, including light, would be confined to that surface (like trains on their tracks), but gravity itself would be free to leak along the extra dimension, explaining why it behaves differently. One of the most interesting proposals was the idea that the evolution of the Universe, as we see it throughout cosmic history, could be the result of this brane's motion along the extra dimension.

While I was working on my PhD, I was fortunate to work on particular realizations of these ideas with Neil Turok and Anne Davis before joining the group of Robert Brandenberger at McGill University. The three of them had some of the most original and inspiring ideas for the origin of the Universe and for why we seem to be observing only three dimensions of space. I was then hired at McMaster University and the Perimeter Institute for Theoretical Physics by two other incredible physicists, Justin Khoury and Cliff Burgess. Together, they taught me almost everything important I know about gravity and how to embed it within a field theory framework. In this, we were strongly influenced by the train of thought of Nobel laureate Steven Weinberg, who sadly passed away in 2021. From models of "supersymmetric large extra dimensions" (SLED), pioneered by Cliff and collaborators, to the model of "degravitation" led by Justin and his collaborators, these were fun and exciting times when the scientific community was not afraid to explore every possible aspect of gravity and its interconnection at the particle level to make sense of our Universe. Before the discovery of the Higgs, we even proposed a model where the Higgs could hold a special place within the extra dimension, in the hope of addressing the hierarchy problem.

This heady time also coincided with the end of my dream to become an astronaut. Once I awoke, it was time to fully embrace my research career. I knew that if I wanted to make it, I would need to further develop and prove my independence, so I decided to start working on a new extra-dimensional model I had concocted in the hopes of achieving some of the features of degravitation that Justin had previously proposed, while heeding the lessons learned from the SLED breakthroughs developed by Cliff and collaborators. After working on the project for about six months, I had the opportunity in 2009 to present my unpublished results at the annual International Conference on Particle Physics and Cosmology (COSMO), one of the largest conferences in our field, which was taking place at CERN (Conseil Européen pour la Recherche Nucléaire) in Geneva. I was excited to present my work, but I was also looking forward to listening to some of the speakers at the conference, including my partner Andrew Tolley and Gregory Gabadadze of New York University (NYU), both of whom were giving plenary talks back-to-back on the final day of the conference (figure 6.2).

The evening before their talks, I decided to look over the program for the next day to see what else was on offer. In the initial program, Gregory's talk had been listed as TBA (to be announced), but it had recently been updated and, to my surprise, sounded strangely familiar. Andrew and I dashed to the arXiv, a website where scientific articles are posted, to see if there were any clues about his upcoming talk. My heart stopped. We realized that his most recent scientific work posted on the arXiv site proposed an identical model to the one on which I had been working for the past several months, and he had already developed it with much greater depth and clarity. I had been scooped!

That was not an entirely new feeling. The experience of being scooped is common in academia. When it happens there is usually little else to do but to swallow your ego and try again, better

FIGURE 6.2. Colleagues Gregory Gabadadze (*top*) and Andrew Tolley (*bottom*), with whom I developed our theory of massive gravity. *Source:* Simons Foundation (photo of Gregory Gabadadze).

and faster the next time. In fact, I was once beaten by only a few hours, hours that would have been seconds if it hadn't been for the spotty Wi-Fi connection at the Geneva airport where I was waiting for my flight to attend a conference. Those few hours ended up being significant in terms of how much attention and consideration our results received. But, once again, the beauty of falling is not in the fall itself, but in how you get back up.

On this occasion, Gregory had clearly derived the results first. He had scientific priority, and it would have been easy for him to ignore my work. But as anyone who knows him can surely confirm, Gregory is not your typical scientist. Rather than dismissing my work as a simple echo of his (which it was, albeit unknowingly), Gregory took the opposite approach. Within hours, he had invited me to join what turned out to be the most successful collaboration of my career.

I mentioned earlier that when pursuing hard calculations, I sometimes pause my work in the middle of a calculation, resting on a high note for the evening and avoiding the possibility that an unforeseen complication will result in a night of restless sleep. Gregory does not abide by this approach, and that quickly proved to be the right attitude. Once our collaboration began, there was simply no messing about. When I would visit NYU, no distractions would take us away from the blackboard, where we would carry out all our thoughts and ideas, calculations and diagrams. When we thought we had finally reached a result and had an idea about how to proceed, we wouldn't congratulate ourselves and schedule a meeting for some point in the future. Instead, we simply had to know the answer straight away and would work all day and all night, and the day that followed, day after day, until the last possible minute before I had to catch my flight back home. And that's how we made progress. By the end of my visit at NYU, we had developed our first model of extra-dimensional gravity that possessed all the qualities we expected of it, at least to the order we could scrutinize.

Massive gravity was still buried beneath layers of no-go theorems endorsed on the one side by the most prominent experts on gravity, and on the other by leaders in quantum field theory. They all contended that the graviton could not be massive. In fact, I was so convinced by those theorems that when I was interviewing for faculty positions around the same time, I made a point of emphasizing what appeared to be well-founded no-go statements: a purely four-dimensional theory of massive gravity could not exist without ghosts, which is why extra-dimensional alternatives were the best hope. I had no idea how wrong I was at the time.

It was during one of these campus visits, where the faculty interview involved two full days of intense exchanges with members of the physics department, that I began to realize something wasn't quite right. When I returned to the bed & breakfast where I was staying, the thoughts had been running in the back of my mind for long enough that I started feeling very uncomfortable about some aspects of my work. I couldn't put my finger on it at first, and for weeks I was convinced I must have made a mistake. I frantically ran through my calculations, always coming up with the same result: somehow, the model Gregory and I had created seemed to be avoiding the plethora of no-go theorems that forbade massive gravity. Our model did rely on extra dimensions in some ways, but it could still be interpreted entirely in four-dimensional terms, so the no-go theorems should have applied. The ghostly sixth mode that plagued any theory of massive gravity should have been present, and yet in the calculations we had performed so far it never showed up [20, 21].

Tickling an Elephant

The Blind Men and the Elephant
It was six men of Indostan
To learning much inclined,
Who went to see the Elephant
(Though all of them were blind),
That each by observation
Might satisfy his mind.

BY JOHN GODFREY SAXE, 1872
(based on earlier parables dating as far back as 500 BC,
including the Buddhist text *Paṭhamanānātitthiyasutta*)

"The Blind Men and the Elephant," a poem by John Godfrey Saxe, retells the ancient Indian legend of six blind observers (who could well have been women) who learn and imagine what an elephant is like through touch. It is a tale that dates back more than 2,500 years and recasts with infinite wisdom the process of scientific exploration. Exploring the polarizations present in a theory of massive gravity is very much like exploring the facets of a mysterious beast. The various types of glight polarizations that can be produced in a gravitational theory don't just knock on your door, politely introducing themselves with their name and title. Instead, you must tickle the beast, and if you've done it right and gently enough, it may present one of its limbs.

Our job, as the blind observers of this world, is then to analyze that limb and figure out what it might be. Could it be the hind leg, common to most mammals, analogous to the + and × modes already present in general relativity? Could it be one of the front legs, which distinguish the elephant from, say, a kangaroo, just as the existence of the ± 1 modes set a theory of gravity apart from many of its close companions? Could it be the tail—or this breathing mode, shared by many theories of gravity, albeit not by general relativity? Or is it the trunk, the elephant's most unique feature—in our case, the infamous ghost?

In the dark, the various limbs are difficult to identify, and the best one can usually do is simply count how many different extremities the beast can wiggle and jiggle as you tickle her. Over the past eighty years, when scientists have probed massive gravity, they have consistently identified all four legs, the tail, and the trunk—the six polarizations I have described above. Despite this, the model we devised seemed to have only five—it carried no ghost. How could it be? Where was our elephant's trunk?

That's when it hit us: what if all six modes are present, but they aren't able to move independently? When it comes to your average elephant, she can keep her trunk still while the tail wags, or keep her tail still while the trunk reaches out for a pea-nut. Imagine, however, that the trunk of the beast was locked to its tail or attached to the leg, unable to move on its own. In this case, it wouldn't really count as an independent limb. That's ex-actly what was going on in our model, and what had been elud-ing the entire scientific community for so long. Massive gravity may appear to have six limbs, at least when palpated, but not all of them are free to move at will. One can get stuck, constrained, unable to move freely. If the ghost was always attached to an-other limb, unable to wander freely, it would never spook us. It would never be a polarization in its own right.

What is even more remarkable was that there wasn't just one proof of massive gravity's irrevocable ghost. There were at least four different types of arguments, each of which seemed to in-dependently convince everyone that a theory of massive gravity without a ghost simply could not exist. But once we realized how simple it was to restrain this ghost, our motivation for delving deeper into the intricate depths of the other no-go theorems grew much stronger. That's when we realized that many of these arguments were either not entirely fleshed-out or not as independent as people had thought; they each relied on a similar type of counting that ignored how the various modes could be constrained. To our surprise, some of them simply had errors in them. They were subtle ones, of course—errors that would not have made a difference on their own but, when multiplied, changed the conclusions entirely. Slowly, brick by brick, we were able to dismantle the four-story-high wall that stood in the way of reaching a consistent theory of massive gravity.

Now that we had dismissed these initially insurmountable arguments and discovered the rules that we needed to follow to avoid the ghost of massive gravity, we "simply" had to demonstrate how a full-fledged beast could emerge: a four-legged beast with a tail, making sure that no trunk was present or at least not free to move. I had just started a fixed-term assistant professorship at the University of Geneva at the time, while Andrew was still based at the Perimeter Institute for Theoretical Physics in Canada. During his visits to Geneva, though he was always welcome in my department, Andrew always seemed to prefer working in a boulangerie in the city center. Perhaps the famous *ramequin au fromage* and *delices aux beurres* had an appeal that my office could not quite compete with. So that's where I found him one late afternoon as I was walking back home. I took a seat at his table, and we began to discuss the massive gravity beast on which Gregory and I were working.

People often adopt a bemused expression when they hear that Andrew and I work on similar topics and have coauthored multiple papers together. As it happens, for the first few years of our relationship we rarely discussed physics. But as time passed, it became increasingly difficult to resist sharing all our joys and excitements, our ups and downs, especially when one of us seemed to be so close to something special and magical without yet quite grasping it; that's often when Andrew's incredible insight has completely shifted my perspective in a matter of seconds. Mixing professional and intimate relationships is not something entirely trivial, though. It took years to make it work, and even today we need to be perpetually mindful about the boundaries. However, it has at times been helpful in our scientific brainstorming discussions. Even in the midst of a heated debate with colleagues, the questions and challenges remain civil and polite. When you discuss research with your partner, it is not that the discussions become antagonistic, but an entire barrier of protection falls. You cannot hide behind politeness to conceal what you may not know, and no one holds back in pressing an evasive answer. All that matters is reaching the right answer and discovering what nature makes possible.

Within a few minutes of talking with Andrew in that cozy boulangerie, the first hints of how to build the full theory of massive gravity became clear. They had, in fact, been present all along in the model Gregory and I had developed. All we had to do to describe a sensible theory of massive gravity was use the right variables to build the interactions. It is a common experience in physics that what initially appear to be long and complicated arguments end up being almost trivial when the right tools are used. In Einstein's theory, the main object that is used to describe the geometry of spacetime is called the metric. The metric is a set of ten numbers that tell us how to measure distances in space and time at a given point. As we move around in space and time, these ten numbers change, and this encodes how space and time expand and contract in different ways, describing the curvature of spacetime itself. In massive gravity spacetime is also described by this same ten-component metric, but we have to use special bricks to ensure that our model remains free of ghosts. These bricks appear horrendously complicated, and we could not discern an underlying structure to put them together.

Instead, taking inspiration from the way general relativity operates in higher dimensions, we became convinced that a new type of brick could be used, one that played the same role as an object known as the extrinsic curvature. Extrinsic curvature describes how a surface is bent or deformed in an otherwise flat space, and just like the metric, it has ten components. In the higher dimensional models that Gregory and I had developed and discussed at CERN, the graviton mass arose from the extrinsic curvature. While the ghosts seemed to be avoided at low orders, we knew by then that this was not quite the final answer. Nevertheless, we were increasingly convinced that by finding the right analogue for the extrinsic curvature, it should be possible to engineer a theory of massive gravity in four dimensions, free of any of the pathologies that had until then been prophesied.

A couple of days later, on his way to the gym in Geneva (no doubt to work off those ramequin au fromage and delices aux beurres), Andrew realized that if the interactions were built out of an apparently unusual new brick, one that took the form of a square root function of the metric, which resembles the extrinsic curvature in some limit, the properties that Gregory and I knew had to be true would follow automatically. He cycled to the train station to tell me about his insight, just as I was about to board a train for Paris, where I had been invited to give a seminar. I was used to these three-hour train rides from the time I used to study in Paris, but on that day the journey passed in the blink of an eye as I frantically calculated the effects of this square root brick. When I arrived in Paris at the Gare de Lyon, I immediately called Andrew. Everything was right. We had our answer [22].

In this square root structure, we had a perfect analogue for our extrinsic curvature, and everything was working like clockwork. We spoke with Gregory as soon as we could, and for the first time in my life all the pieces of the puzzle fell into place with incredible ease. After a week, I had pages of calculations, including contributions to twentieth order in perturbations, involving thousands of terms that, once the right structure was applied, would miraculously prevent the ghost from ever appearing. After sixty years of wandering in the wilderness, massive gravity emerged in its full glory. By combining Gregory's and my calculations with Andrew's square root structure, we knew not only why the previous no-go arguments against massive gravity had been wrong but also how to bring the full theory to life.

Graffiti in the Sky

At this point, I'm at a loss for analogies and comparisons to describe how our theory looks, smells, and feels. The idea of tickling an elephant can only take you so far, at which point it's time to turn to the mathematical representation of general relativity and massive gravity.

One equation we have seen repeatedly is Einstein's infamous $E = mc^2$, which follows from special relativity. It tells us how to relate inertial masses and energy. Einstein's general theory of relativity, on the other hand, tells us how the curvature of spacetime is related to the energy density and pressure of everything living in its web. Einstein's field equations of general relativity are now so trendy that they can even be found as funky graffiti

FIGURE 6.3. Graffiti representing Einstein's equations of general relativity.
Source: Photograph by bbuong (iStock).

artwork lighting up an old rusty locomotive in Bolivia, as shown in figure 6.3.

This may appear as a complicated equation with many Greek indices capturing the various directions in space and time, but in essence it tells a very simple story: the energy density of everything in the Universe (captured by "T") feeds into the curvature ("R") of spacetime. If the quantum vacuum energy of particle physics carries a large energy density, T is then large, and this directly feeds into a large curvature, which in turn produces significant cosmic acceleration. According to general relativity, our spacetime should be so curved that our observable Universe would be no more than a centimeter in size. The rest of the symbols we see in figure 6.3 are the constants we have already introduced. The letter "G" is the Newtonian constant of gravitation, and "c" is the speed of light in the vacuum, both of which are

FIGURE 6.4. Our theory of massive gravity. We like to call it "ghost-free," though the community has called it dRGT, after my work with Gregory Gabadadze and Andrew Tolley. The key difference between massive gravity and general relativity is the inertial mass of the graviton, which enters with the letter "m" on the top right corner. The main complication lies in finding the proper mass function "U" that exorcises the ghost. This is what my daughters are hard at work deriving on the chalkboard.

key to determining how much an element of mass or energy will affect the local spacetime curvature in which it lives. All of these quantities are related to the Planck energy scale "M_{Pl}" that is introduced in the next drawing.

I haven't had the chance to draw my own graffiti on a Bolivian locomotive yet, so the best analogue I have for our theory of massive gravity is that drawn by my four- and six-year-old daughters a few years ago, as depicted in figure 6.4. (Unfortunately, my two-year-old was unable to reach the board at the time, which is probably one of the few valid reasons for not wanting to join the fun.)

This may seem rather abstract, but together these obscure symbols tell a simple story of their own. This story is similar to that whispered by general relativity but with its own little twist: the spacetime curvature "R" is affected by the energy density of everything in the Universe *and* by the graviton mass herself, including the additional polarizations offered by a massive graviton, which are hidden in the function "U." Einstein told us that energy is mass and that curvature is related to energy density. In massive gravity, all of these elements are combined into one overarching theme: the energy density of all the bits and bobs living in our Universe, together with gravity's own mass, feed into the spacetime. Phrased in this way, this seems like a straightforward generalization of general relativity, with one key difference. It is now possible to have a large quantum vacuum energy without necessarily implying that the Universe is curved to excruciating levels. The graviton mass, or rather the additional flavors of gravity hidden in the graviton mass, can act as an airbag or safety mechanism, absorbing the shock (more precisely, the inflow) of vacuum energy. Instead of curling the fabric of spacetime to ridiculously small scales, the vacuum energy and the graviton mass compensate for one another, leaving the standard curvature of our Universe virtually unaffected by the existence of vacuum energy. Rather than being an unwanted side effect of massive gravity, the additional polarizations that come along with the graviton mass make our Universe a viable place after all.

But being able to absorb a fantastically huge vacuum energy is only one aspect of the story. The overall implications of this new theory of massive gravity are rich and profound, as we will explore in chapter 7. However, before doing so, there is one last part of the story we should address: the beginning itself. Cosmologists agree that 13.8 billion years ago, near the very outset of the Universe, the spacetime curvature was much (much!)

higher than it is today. So high, in fact, that the mass of the graviton would have been negligible. How then, could the graviton mass have had an effect? The resolution is intriguing in its simplicity. Any theory of gravity is a theory of the intertwined nature of space and time. If massive gravity affects how gravity behaves over very large distances, it should also affect how gravity behaves over very long timescales.

As applied to our tiny Universe in its infancy, graviton mass has no effect. But over sufficiently long periods of time, on timescales as long as the age of the Universe today, the graviton mass kicks in and relaxes the large effect of vacuum energy. In practice, it will take an infinite amount of time for the graviton to wash out the vacuum entirely. This is why, at present, though most of the vacuum energy has already been absorbed by the graviton mass, there is still a tiny residual effect. That effect is precisely what we observe today and explains the small rate of accelerated expansion of our Universe. At least, this is the tale that the graffiti of massive gravity would whisper in our ears if we tried to decipher its symbols.

These symbols may appear dark and alienating, but the concepts they convey are ultimately ones we are very familiar with—ones that are fun at heart and tell us that there is a limit to everything, including gravity. Giving it tangibility could help us reconcile the vacuum energy of particle physics with the acceleration of the Universe, harmonizing the two pillars of modern physics. Humanity has been hunting for a tangible understanding of our Universe for decades. As if to humor us, a new possibility has emerged, potentially bringing us a step closer to understanding our story, that of our Universe's origin and its fate.

CHAPTER 7

Our Lives in Pursuit
of Gravity

My personal quest to become an astronaut involved years of commitment. In retrospect, perhaps all that time and energy would have been better spent on more certain or constructive pursuits. But I don't regret a second of it. Learning to fly, for instance, was a life-changing experience. Like becoming a physicist, becoming a pilot wasn't exactly a walk in the park. I had to rise before the sun to log my flight hours, and I spent a considerable amount of time preparing for my pilot license exams.

There was, however, one aspect of this process that I relished: at no point was there any pressure on me to do anything original or unique. At least for the basic single-engine license, the only real requirement is to calmly follow the correct procedure at the right time for the right reasons. Since I had practiced each procedure hundreds of times, I had the confidence I could do it again. I didn't have to imagine or dare something new, nor prove that the way I performed a specific maneuver was the best or only possible option; I wasn't even required to establish that I was the best pilot by any particular metric. All I had to do was apply the appropriate method without making mistakes. The idea that the absence of failure is sufficient for success is so comforting that I frequently wish I could return to it. To be an astronaut or a theoretical physicist, not failing is never enough. In fact, failure is deeply embedded within the process of scientific research. As for our theory of massive gravity, not failing was never the goal; it was merely the start of our journey.

From distances as small as the thickness of a piece of paper to as large as the span of galaxies, general relativity has proven itself an unfailing guide, achieving unparalleled agreement between theory and real-world observations. For a challenger to dethrone general relativity, it is not enough merely to show that the theory achieves similar results. No, it must possess distinguishing features that set it apart and show that it outperforms general relativity in certain ways. Since all observations and experiments to date have been in complete agreement with general relativity, these signatures must be subtle or small enough to explain the fact that we have not yet noticed them, yet significant enough to differentiate the challenger from general relativity in the future.

Just as a new QuantiFERON tuberculosis test was introduced shortly before my astronaut selection, new theoretical, observational, and experimental missions are continuously being devised to test our models in deeper and more exhaustive ways. To nonexperts, coming up with a new theory may appear to be a great accomplishment. But the true challenge is surviving what comes next. The rigor of the tests to which I was subjected during the astronaut selection process pales in comparison to the scrutiny that our model of massive gravity has faced, and continues to face, from the scientific community. This scrutiny began in earnest once Gregory Gabadadze and I had finalized our second set of results and posted our paper on arXiv, and I began the standard journal submission process.[1]

Because our results revealed for the first time the caveats hidden in all "no-go" claims against massive gravity, it seemed appropriate to submit our manuscript to the same journal that had published one of the strongest no-go statements in recent years. How naive of me! Naturally, I never expected the peer-review process to go smoothly. After all, scientific results are meaningless if they are not dissected and reanalyzed by other experts. I was prepared for our paper to be met with some skepticism and resistance, as well as for the potentially long discussions that would ensue as we responded to the referee's objections. Yet I was apparently overly optimistic: the editor of the journal simply dismissed the manuscript before it even reached the refereeing stage. The editor's main concern, as far as I could tell, had less to do with the scientific rigor and validity of the work and more to do with ensuring that no challenges were raised against their colleagues and ideology.

Fortunately, our manuscript was quickly accepted by an equally reputable journal [21]. Within a few years, together with [22], our findings came to be recognized and ranked among the most influential discoveries in physics during the past decade. However, as I soon discovered, the initial editor's reaction was a bellwether for the reaction of many members of our community. Our work had clearly touched a nerve. After initially dismissing our results, some scientists claimed that they were straightforwardly incorrect, while others claimed that they or their colleagues had already derived them. To my great amusement,

1. The dissemination of scientific research now occurs almost entirely through the free online arXiv, which anyone can access, but publication in specialized peer-reviewed scientific journals is still necessary to advance one's academic career.

certain scientists made *both* claims, though I still don't understand why anyone would be so eager to claim that they had derived such obviously flawed results, but that shall remain one of humanity's small mysteries.

To the nonscientist, it may not be obvious how other scientists could so readily dismiss our results. After all, in our published work we made sure to show how to prove our results using mathematics, which many regard as the ultimate language, encompassing both logic and truth. This is undoubtedly one of the factors that drew me to it; I could find truth and comfort in the laws of nature that cut through any level of miscommunication. But it doesn't take long to realize that things are rarely so simple. Those who focused on the underlying physical aspects of our model or the accompanying mathematical proof tended to agree with us. Soon enough, the logic behind our proof was generalized by Rachel Rosen and Fawad Hassan, and then by Rachel and Kurt Hinterbichler. Others quickly followed. The math checked out. The skeptics' criticism, on the other hand, had nothing to do with the logic of our proof; rather, their skepticism almost always stemmed from differences between the mathematical language we used and that to which they were accustomed.

There are perhaps as many cultural differences in the mathematical language we use, or the style of scientific argument we prefer, as there are between people who happen to live within the same society. In the scientific case, these differences are not primarily geographical; rather, they simply reflect the richness and breadth of the directions in which physics and mathematics have evolved. If this difference in scientific languages seems abstract or even counterintuitive, think of the real number π. This is one of the most important numbers is mathematics, which we all know is roughly equal to 3.14. We can reach this conclusion in different ways:

- Some will draw a perfect circle of radius R on a flat piece of paper, determine its circumference C with a soft tape measure, and use the formula $C = 2\pi R$ to infer the value for π.
- If you don't have a tape measure, you could instead fill a perfectly spherically symmetric balloon of 1 cm radius with water and weigh it. Knowing that the volume of the balloon is $4\pi/3$ cm^3 and that a cm^3 of water weighs 1 g, π will be given by three-quarters of the mass of the balloon in grams.
- Others will have learned the formal Gregory–Leibniz series rule, which tells us that $\pi/4 = 1 - 1/3 + 1/5 - 1/7 + \ldots$ where the series continues forever, alternating between adding and subtracting fractions of consecutive odd numbers.

In a perfect (flat) world, each of these techniques is equally valid and will lead to the same result within some level of precision. But imagine you were in a curved space. In that case, the piece of paper would not be exactly flat, and the formula for the circumference of a circle would have to be adapted. We would need to rethink it. Inferring π from the mass of a balloon would also need to be adapted, but in a different way. If the mathematical Gregory–Leibniz series rule gives us a value for π of 3.141592653589793 ... then all other methods *must* give the same value. But for someone unfamiliar with the Gregory–Leibniz series rule, attempting to infer the value of π using the circumference method would yield the wrong result. Until someone comes along and explains how to account for curvature, that person may be inclined to doubt the mathematical Gregory–Leibniz series rule, or even dismiss it altogether.

To be sure, mathematics does not lie. If something is proven in one specific mathematical language, it must be true in all possible languages; we just need to properly understand the underlying foundations. So, over the years, every time a new colleague had a different perspective, we were compelled to learn their preferred language, translating our logic in different terms and adapting our thoughts to this new mathematical culture. With the help of our colleagues, we gradually translated our underlying argument into multiple mathematical and physical representations. They all came with different names and variants, from the so-called Hamiltonian constraint, to the Stückelberg constraint, to the helicity constraint, to both the constrained vielbein and unconstrained vielbein, and to other "forms." These languages all rely on different methods to count the number of polarizations and determine whether the ghost is present; just like in calculating π, unless the foundations of the method are properly understood, they can easily yield incorrect results. For five years we had to repeat this process until no unanswered questions remained. Our claims about massive gravity were judged not by a closed jury of twelve peers but openly by the entire multicultural scientific community.

This inquisition felt overwhelming at times. Beyond the well-meaning yet intense scientific scrutiny, I received ominous messages threatening that if I didn't retract my work I would face serious professional consequences. Some messages contained even more personal insults. Yet, through it all, the confidence that our results were mathematically correct kept me going, and the process itself afforded me an incredible opportunity to learn from our community's rich diversity of languages and techniques.

Having deflected most of the arguments against our theory, or at least established the absence of the ghost, the most exciting part of the adventure could finally begin. While massive gravity had not yet been "proven correct"—no theory, not even general relativity, is ever proven right in this respect—the fact that it hadn't yet failed so far meant that we could imagine testing it against reality, letting nature determine whether this alternative to Albert Einstein's long-standing theory of gravity would hold up.

Lightness of Gravity

For massive gravity to make sense at all, the mass of the graviton should be very small, with a range close to the observable Universe today, if not larger. In terms of energy scales, this corresponds to a mass of about 10^{-32} eV or smaller, or close to thirty orders of magnitude lighter than the lightest massive particles we know, the neutrinos. In the context of dark matter, many other models include massive particles with very light masses, but a mass of 10^{-32} eV would be far below anything else expected in nature. So the first question we should ask ourselves is whether such a small mass is even reasonable to consider. Shouldn't we expect the mass of other particles to rub off on the graviton and increase its value? This question is, in fact, at the core of the new cosmological constant problem, and answering it requires understanding the details that come into play when constructing a theory of massive gravity. Now that we have such a theory at hand, we have been able to follow the interplay between massive particles and the graviton mass and confirm that the lightness of the graviton is, in fact, reasonable (or what we call "technically natural").

One way to understand why the graviton can be much lighter than every other massive particle is to go back to general relativity, where the graviton is not only lighter, but massless. In general relativity, the absence of mass is dictated by the equivalence principle, which forces gravity to interact with anything and everyone in precisely the same way. Gravity is then so rigid that nothing can make it change behavior, and other massive particles cannot rub off their mass on the graviton: once massless, always massless.

Now, in massive gravity the rigidity of gravity and the equivalence principle are slightly broken. This is precisely what allows us to make do with a large vacuum energy, which would otherwise cause the Universe to curl to less than a centimeter in size. But those principles are only broken by an amount that depends itself on the mass of the graviton. If the graviton is very light, the principles are only very weakly broken. In that case, while other massive particles can, in principle, slightly rub off their mass on the graviton and make it slightly heavier, in practice this effect is suppressed by the graviton mass itself and, at least in theory, is never significant. But to really understand what happens in practice, we need to compare the theory with some actual observations.

(Space)Time Will Tell

There is no limit to the number of tests or health checks to which a new physical theory can be subjected. Fortunately, for many conditions there are early warning signs that can help us diagnose whether we expect the model to be compatible with what we observe and experience. The main motivation behind massive gravity was to relax or weaken the effect of gravity on large cosmological distances, so once we had developed our model, we immediately began investigating how well it actually performed on this front.

This investigation (and many others) would not have been possible without the ingenuity of David Pirtskhalava, the brilliant creativity and determination of Lavinia Heisenberg, and the deep perceptive intuition of Andrew Matas. David was one of Gregory's PhD students, and both Lavinia and Andrew began their PhD work with me around this time. Our ongoing conversation, as well as their passion for the project, kept me going during the period of intense confrontation with the rest of the scientific community. Moreover, as we began the second stage of our massive gravity adventure, they brought more insight to the project than many of the senior figures with whom we discussed our work.

In order to test whether massive gravity did, in fact, weaken the effect of gravity over large cosmological distances, we needed to make gravity "massive" without making it so inert that it would relax too quickly and disturb the whole structure and evolution of the Universe. While it is okay for gravity to switch off at distances comparable to our current observable Universe (about 13.8 billion light years)—in fact, we needed it to do so— if this happened any sooner, signs of this relaxation would have already been observed. Indeed, if the graviton were too massive, it would not be possible for the clusters of galaxies millions of light years apart to be gravitationally bound. That observation alone provides one of the strongest bounds on the graviton mass, constraining it to be less than the inverse of about a million years, which corresponds to about 10^{-29} eV in energy scale. Given that we had in mind a mass on the order of 10^{-32} eV, we were still safely within that bound. Nevertheless, it is remarkable that the presence of gravitationally bound clusters of galaxies places some of the most stringent constraints on the mass of the graviton.

In addition to our observations of intertwined clusters of galaxies, we have an incredible number of resources to test gravity. One of the most important is the direct detection of glight or gravitational waves emitted by the merger of two black holes or neutron stars. Far from being a single musical note played repeatedly, the gravitational waves observed by the U.S. Laser Interferometer Gravitational-Wave Observatory (LIGO) and European Virgo observatory teams on Earth are a symphony of notes played in precise harmony. As the two orbiting black holes or neutron stars become more intimate in their dance, the two celestial objects emit notes of increasing pitch (glight rays of bluer colors). These signals frantically increase their frequency all the way up to the merger.

One of the features of massive gravity is that, as with the propagation of any massive wave, the mass affects various colors or frequencies differently. High-frequency waves, which are emitted later in the symphony, travel at close to the speed of light, virtually unaffected by the presence of a small mass. Earlier waves, on the other hand, have a lower frequency and are more burdened by their mass, which slows them down as they travel across the Universe. If gravitational waves are, in fact, massive, the difference in speed between the low-frequency modes emitted earlier and the high-frequency modes emitted later would be noticeable, and the signal would appear distorted.

Though glight was only detected for the first time a few years ago, the gravitational wave signal emitted by black hole mergers has already been imprinted on nearly every type of clothing and accessory imaginable, including caps, face masks, ties, cufflinks, dresses, phone cases, and earrings. A simplified version of that signal is represented at the top of figure 7.1. If all the colors of glight traveled at precisely the same speed, the shape of the signal we would receive would be the same as that emitted. This would be true if gravity were massless (as in general relativity) or if its mass were sufficiently small. If gravity were "very massive," the signal would appear distorted because higher frequency waves emitted later in the process would

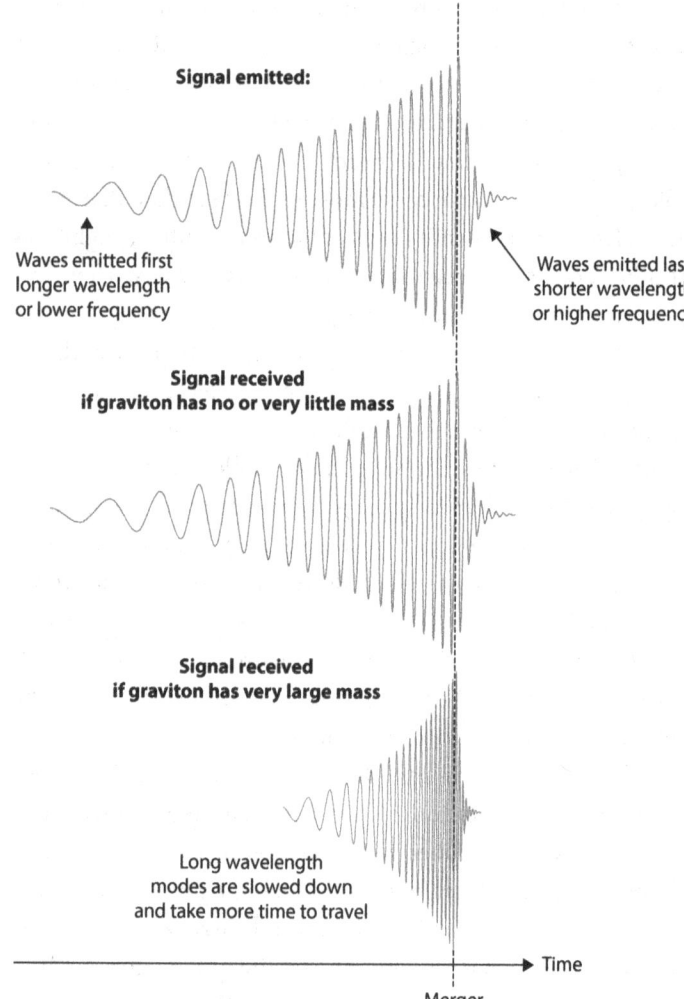

Signal emitted:

Waves emitted first
longer wavelength
or lower frequency

Waves emitted last
shorter wavelength
or higher frequency

**Signal received
if graviton has no or very little mass**

**Signal received
if graviton has very large mass**

Long wavelength
modes are slowed down
and take more time to travel

Time

Merger

FIGURE 7.1. Difference in the gravitational wave signal we would detect if the graviton were massless (as in general relativity) or massive.

catch up with the slower, lower-frequency waves, compressing the resulting signal that we receive.

For realistic values of the graviton mass, the difference between the signals would be impossible to notice with the naked eye. To allow us to visualize these effects, figure 7.1 shows a signal with a "dramatically" enhanced graviton mass, about twenty orders of magnitude larger than we would expect. By carefully analyzing the observed signal and comparing it with our expectations about the emitted signals, the LIGO and Virgo teams have been able to put a solid upper bound on the graviton mass, constraining it to be smaller than 10^{-21} eV. Because our theory of massive gravity puts the graviton's mass somewhere in the neighborhood of 10^{-32} eV, our theory easily satisfies this constraint. As an aside, it's worth noting that when it comes to light, the mass of the photon is only constrained to be less than about 10^{-20} eV, which is a weaker bound. Even though we've only been detecting glight for a few years, and even though we had to overcome significant infrastructural challenges in order to detect glight, we already know more about the mass of the graviton than we do about the photon! The fact that we have surpassed thousands of years of ordinary light detection with just one observation is a testament to how far we have come in our understanding of gravity.

Using the detection of multiple new glight signals, some of which were accompanied by light, scientists have recently refined the constraints on the mass of gravity to be less than about 10^{-22} eV. Although this is still much less restrictive than the bounds established by the existence of gravity-bound galaxy clusters, future research in this area is quite promising. The next decade will witness gravitational wave observatories set in space, a project currently underway with the Laser Interferometer Space Antenna (LISA) missions. The arms of LISA will be over two million km long, following Earth as it orbits the Sun. And why stop there? With the ongoing Pulsar Timing Array (PTA) missions and their detections of a low-frequency background of gravitational waves, we are entering an era in which a network of pulsating stars in the sky will act as our gravitational wave observatory [3]. We are literally using stars as our instruments, allowing us to observe glight with lower frequencies than ever imagined and probe the nature of gravity even further.

Pi in the Sky

As the entire point of massive gravity is to change the behavior of gravity over large distances and long timescales, the best constraints on its mass are likely to come from looking at the largest of scales. If gravity is massive, no matter how small its mass, there will always come a point where the effects of its mass will show up. Perhaps the best way to detect it is simply to wait a few billion years to see if the Universe follows the pattern dictated by a massive or a massless graviton. Were our life spans (and patience and dedication) as eternal as our Universe, such reasoning would be flawless. Fortunately, for finite beings such as us, there are more practical alternatives.

One possibility is to focus on gravity's various polarizations. One of the reasons that it proved so difficult to develop a theory in which gravity has a mass is that even minor changes to the structure of general relativity can destroy its foundations. As we have seen, the additional polarizations of gravitational waves that appear when gravity is massive give rise to the most serious complications for any alternative to general relativity. In order to develop a viable theory of massive gravity, our primary focus was to ensure that the so-called ghost particles were forbidden from appearing, spontaneously popping out of emptiness to annihilate the entire nature of reality. This was no easy task, but now that we've devised a model that eliminates the ghost for good, we can turn our attention to the other three remaining polarizations that exist in massive gravity, in addition to the two already present in general relativity. Two of them, dubbed the ± 1 modes, are relatively harmless because it is difficult to wake them up and observe their effects. They are dormant and inaccessible in the vast majority of our Universe's environments. The third additional mode, the "breathing mode," on the other hand, has a genuine life of its own. This mode is frequently referred to as "pi" and designated as π because our understanding of it is closely related to a type of subatomic particle known as pions (hence pi), which are responsible for mediating the strong nuclear force inside the nuclei of atoms.

The π field is genuinely gravitational. This mode interacts with nonrelativistic matter, such as stars, planets, dark matter, and so on, but not directly with light or other relativistic particles.[2] This means that if we look at how light bends under gravity, we would get the same result under massive gravity as general relativity (minus tiny corrections due to the graviton's mass) because the π field has no direct effect on light. However, if we consider two massive objects, such as the Moon orbiting around the Earth, the results will differ due to the tiny additional force coming from this new π field. We often refer to this effect as a fifth force because it goes beyond the four known forces (general relativity, electromagnetism, weak nuclear, strong nuclear). The fifth force from the π field generated within our own solar system already provides us with one of the best

2. By "nonrelativistic" matter, we mean massive objects or particles whose speed is small as compared with that of light. "Relativistic" particles have a speed that is close to that of light.

constraints there is on the graviton mass, hence the search for π in our own sky.

If the graviton is close to massless, the π or breathing mode effectively lives its own disconnected life as a hermit with no impact on us. In scientific parlance, we say that the π field is decoupled. The mechanism by which this occurs was first postulated by theoretical physicist Arkady Vainshtein in 1972. It took nearly thirty years to make his argument more precise, but in 2001 Cedric Deffayet, Gia Dvali, Gregory Gabadadze, and Arkady Vainshtein successfully proved how this so-called Vainshtein decoupling could be realized in the context of modifications of gravity arising from extra dimensions; the same arguments apply to our model of massive gravity. They discovered that for small enough values of the graviton mass, this π mode is so busy and caught up in its own games, strongly interacting with itself, that it lacks the strength to bother us. When the graviton is very light, close to massless, the outcome is thus similar to what we have in general relativity. However, when the graviton mass is not completely negligible, the subtle effects of this mode may be noticeable with current observations, producing distinctive signatures.

One potential signature is a correction to the Hulse–Taylor pulsar we encountered earlier. Gracefully orbiting around each other, these two stars shine with even more glight than light, and their loss of energy was the first undeniable confirmation that gravitational waves exist. If gravity were massive, it turns out that glight would come with more flavors than it does under general relativity. This new type of glight, or π wave, allows the stars to shine even brighter (through glight, not through light), which provides, in theory, a way to determine whether or not the graviton has a mass. However, Andrew Matas, Andrew Tolley, Daniel Wesley, and I were able to show that, in practice, this new glow is very faint unless the graviton mass is sufficiently large. For a graviton mass on the order that we have in mind, the effect of this new mode is undetectable via binary pulsars. However, in the Earth–Moon system, though the effect of this π-flavored glight is even smaller, we can measure what happens with such remarkable precision that it provides us with one of the best constraints on the graviton mass. But what exactly does it mean to "measure what happens" in the Earth–Moon system?

During the Apollo 11, 14, and 15 missions, a series of mirrors were installed on the Moon. By shooting laser light from Earth onto these mirrors and observing it reflect back, we can infer the position of the Moon to within 1.1 mm, an accuracy of more than eleven orders of magnitude. This experiment, which goes under the name of the Lunar Laser Ranging (LLR), is one of the most powerful probes of gravity to date.[3] To achieve such precision, one must not only account for the delay of the light as it propagates through our own atmosphere but also the morphology of the Earth as well as the speed of the Moon and the Lorentz contractions it implies.[4] To reach a significantly higher level of precision, we would have to account for incredibly subtle effects, such as the leaves growing on trees in the spring and falling to the ground in the autumn. Even shifts in the climate

3. By now, even more sensitive tests of the equivalence principle have been designed. For instance, the Microscope Space Mission is sensitive to a change in acceleration rate between two atoms, which could, for instance, be caused by a fifth force or an effective difference between their gravitational and inertial mass. Microscope reaches a sensitivity of about 10^{-15}, but the effect of our pi mode in that environment would be harder to notice than in the LLR Earth–Moon system.

4. Since the Moon is moving with respect to the Earth, we must account for special relativity effects that change the notions of distance and time between the Earth and the Moon, in addition to the spacetime curvature. Contractions of lengths and distances due to special relativity effects are called Lorentz contractions.

produce changes in how mass is distributed across the Earth, which affects the precise gravitational pull between the Earth and the Moon.

The LLR mission has provided us with an incredibly accurate picture of the Earth–Moon gravitational pull; so far, no new π-flavor effect has been detected. What we have observed is in complete agreement with general relativity within eleven orders of magnitude precision. So if gravity were massive, and its new π-flavor was in the sky, we would need to ensure that its effect on the Earth–Moon system is at least eleven orders of magnitude smaller than that of its two other cousins (the two + and × polarizations present in general relativity). Assuming you can trust the theory at that level, such considerations provide some of the most robust constraints on the graviton mass to date, forcing it to be less than 10^{-30} eV. This is undeniably small but still within the range of interest for cosmology. It still astounds me that by simply observing the motion of our very own Moon we are able to better understand the fate of the entire Universe, its accelerated expansion, and the sea of quantum fluctuations from every particle known to exist and those not yet uncovered.

The Irrevocable Quantum Nature of Gravity

By this point, I will have undoubtedly offended yet another large portion of our scientific community by arbitrarily switching between the classical notion of gravity, with its gravitational waves, and the quantum notion of a fundamental particle associated with gravity, the graviton, or even worse its virtual counterpart. After all, nearly anyone who has studied advanced physics is familiar with the deep division between the classical world of gravity and general relativity and the quantum realm of particle physics. Reconciling these two realms remains one of physics' greatest challenges. String theory, causal sets, loop quantum gravity, and other nonlocal models of quantum gravity have all been proposed to complete general relativity at high energy scales (or curvature) and unify it with the rest of the quantum world of particle physics. Without going into specifics, although incredible progress has been made over the past fifty years, it is fair to say that we still have some ways to go before we will have a widely accepted fundamental theory of quantum gravity, unifying all the forces of nature.

The theory of massive gravity was never intended to tackle nor resolve this fundamental mystery; however, the fate of massive gravity and that of quantum gravity will always be intertwined. Unlike general relativity and other alternative theories of gravity, which are primarily built with a classical geometric picture of spacetime in mind, the building block of massive gravity is a particle, the graviton, which we consider to be massive. In adopting this point of view, we must therefore embrace its quantum nature from the outset, so our adventure would not be complete without a better appreciation of what it means for gravity to be quantum. While the following investigations are not specific to massive gravity, they are crucial to its development and possible vindication.

In our search for an ultimate theory of quantum gravity (massive or not), we are looking for a theory that will be valid at very high curvature scales, at and possibly beyond the Planck scale. However, if our goal is more modest, and we simply limit ourselves to curvature scales well below the Planck scale, there is no problem in treating gravity quantum mechanically. We refer to this as the quantum effective field description of gravity. And this is why we have not been particularly mindful of the separation between the classical and quantum descriptions of gravity in our discussion thus far.

With our current understanding of gravity on one side and the quantum world on the other, we can even advocate that gravity, like light, *is* quantum in nature. This is true for general relativity and massive gravity alike. The reason for this is that everything is connected and coupled to gravity, including the most elementary particles we know, such as electrons. The quantum nature of electrons is beyond dispute. Yet electrons feel gravity, and in turn gravity is affected by electrons, an effect known as backreaction. If we want to understand *very precisely* how electrons evolve, we need to have insight into the curvature of spacetime in which they live. But even without this knowledge, we know that the presence of those electrons has an effect on the curvature of spacetime. In this sense, electrons "back-react" on themselves by the intermediary of gravity.

Backreaction is present everywhere, in both classical and quantum physics. If you have ever experienced a 360-degree steep turn when flying, you may recall experiencing bumpiness when rolling out from the turn. Bumpiness in the air may seem worrisome, but in this circumstance it is an excellent sign—a sign that the aircraft drew a perfectly level circle in the air, rolling out of the turn precisely where it started. When the aircraft enters the turn, it shakes the airflow, which in turn jiggles the aircraft as it rolls out. In other words, the bumpiness is caused by the aircraft's backreaction.

Of course, electrons are not small gliding planes, and gravity is not a medium like the air. Nevertheless, a similar idea applies. Electrons can affect themselves through the intermediaries of light and gravity. Yet electrons are also quantum in nature. No matter the precision of our instruments, we cannot determine both their position and momentum with arbitrary precision. And when orbiting around the nucleus of an atom, electrons cannot be wherever they wish: they obey rules set by quantum probabilities. If electrons were able to backreact on themselves through the intermediary of a classical system, we would be able to measure this classical system with arbitrary precision and infer precisely the position and momentum of the electron in a way forbidden by quantum physics. The quantum laws of probability would then simply fall apart. They would "break unitarity," meaning that the probabilities for some processes could be negative or even complex (not real). There could be processes with a greater than 100% chance of occurring. Everything, including nature itself and the structure of reality, would be very different. To preserve unitarity for a quantum electron, gravity, light, and anything else it couples to must follow the same quantum rules. From that perspective, the quantum nature of gravity is not merely a possibility, it is an absolute necessity, without which the laws of nature as we know them would crumble. This is true for general relativity and unavoidable for massive gravity.

From Waves to Particles

The distinction between the more classical concept of a wave and the quantum concept of a particle is a family matter at heart. The first fundamental particle, the electron, was discovered in 1897 by Sir Joseph John Thomson, earning him the Nobel Prize in Physics in 1906, when his son, George Paget Thomson, was a teenager. The Nobel Prize for discovering the particle properties of the electron had already been awarded to his father, so the only option left for George Thomson was to prove instead the wave-like properties of the electron, a discovery that earned him a Nobel Prize of his own in 1937 (together with physicist Clinton Davisson, who independently had made the same discovery). By then, George Thomson was a professor at Imperial College, and his portrait still hangs in the entrance of the physics department, reminding me of the fundamental wave–particle duality of nature every time I enter the building.

The central idea behind Thomson and Davisson's discovery was to show that the electron could be diffracted. Diffraction is a phenomenon in which waves bend when they encounter an obstacle or an opening. Water wave diffraction is easily observed in a canal or swimming pool, and light (and almost certainly glight) in the form of a wave exhibits the same property. Thomson and Davisson's breakthrough proved that "something" as fundamental as the electron could enjoy a dual description, acting as both a particle and a wave. This wave–particle duality had been proposed in the 1920s by Louis Victor Pierre Raymond, 7th Duc de Broglie, in nothing other than his PhD thesis (no pressure for all subsequent generations of PhD students). The same duality had been predicted by Einstein for light as early as 1905, building on the work of Max Planck, through the concept of "light-quanta," which was at the heart of his description of the photoelectric effect and the main citation for his own Nobel Prize. In proving that electrons also act like waves, Thomson and Davisson confirmed the prediction that particles exhibit wavelike characteristics governed by the rules of quantum mechanics.

Before Einstein and Planck, light was believed to be a purely classical phenomenon of wavelike fluctuations in the electric and magnetic fields. Planck first reluctantly proposed that the energy carried by these waves, which was emitted from radiating atoms, could not be arbitrary and should be quantized in discrete quantities ("quanta") whose energy is determined by the frequency of the radiation. Planck's perspective was that the absorption of light was a classical phenomenon, while the emission was quantized. Einstein extended these ideas with his successful quantum description of the photoelectric effect, which held that these quantized energies were a property of light even in the vacuum. Light of a given frequency could thus only be absorbed and emitted in discrete levels of energy. If you have a laser emitting red light of frequency 455 THz, you will be able to absorb it and gain precisely 0.3 eV, or twice that amount, or even three times that amount, but never anything in between. To gain a certain amount of energy, you need exact change, so to speak. This fundamental quantum of light subsequently became known as the *photon*. For the most part, in our everyday life we can describe the light we receive from our Sun as a wave or a superposition of waves of different colors or frequencies. However, in describing how this light is emitted at the nuclear level in the Sun, we are forced to accept its quantum nature, and it is the particle, the photon, that comes to the fore.

We can, of course, produce light in ways that do not fundamentally make use of its quantum nature, as James Clerk Maxwell was well aware. Accelerating any electric charge is sufficient to produce light in a classical way, a phenomenon that is routinely used in any transmitter and is at the heart of every wireless form of communication, including the signals we receive on our mobile phones. The same classical phenomenon is responsible for the production of the glight that we have observed. Glight is generated by the acceleration of astronomically large masses, such as neutron stars and black holes as they orbit around each other. By observing our first rays of glight, we therefore have gained a firm grasp on the classical aspects of gravity, particularly its wave nature; its quantum nature, however, has yet to be established.

Wave–particle duality is inherent in quantum theory and has been thoroughly tested and established for light, electrons, and all other known particles except one: the graviton. Given the striking similarities we have encountered between electromagnetism and gravity, between electromagnetic waves and gravitational waves—that is, between light and glight—it may seem like it is only a matter of time until scientists prove that gravity also enjoys a quantum particle–level description. Just as the photon is the particle associated with light, the graviton is the particle that should be associated with glight. The graviton should be at the very core of any model of gravity, whether it is general relativity, massive gravity, or any other representation of gravity. Yet exploring the quantum nature of gravity, even at the simplest and most basic low-energy level, will make our previous expeditions look like a walk in the park. Indeed, for our ultimate journey, the finish line is not even in sight. Our goal is not to discover which picture of quantum gravity is ultimately correct (if any), but rather to uncover what any such quest must entail, and what we might learn along the way.

The Ultimate Journey

As we sunbathe on a warm sunny day, our skin typically receives a flux of 10^{17} photons per second per square centimeter from the sun. That's almost a billion billion photons reaching our eyes every second! To prove the quantum nature of light, we must be able to isolate a single photon from among its multibillion companions and show that its energy is proportional to its frequency. This may seem quite daunting; in fact, it was only in 2020 that a group at the École Polytechnique Fédérale de Lausanne was able to develop a megapixel camera capable of capturing single photons. However, the cells in the retinas of some frogs can perceive light at the level of a couple of photons, if not just one. Millions of years of evolution have transformed their retinal cells into ideal quantum detectors.

Detecting a single graviton is, however, a much more difficult task because the gravitational force is nearly a million trillion times weaker than the electroweak force. To understand why gravity is so weak, we can go back to the Planck energy scale, which is related to Isaac Newton's coupling constant and dictates the strength of gravitational interactions as well as disclosing the point at which general relativity has to fail. In energy units, the Planck scale is roughly 100 kWh, or about 10^{19} GeV in particle physics units. For the electroweak force, the equivalent scale is the Fermi scale (named after the Italian physicist Enrico Fermi) or the electroweak scale, corresponding to 246 GeV.[5]

These respective scales differ by over seventeen orders of magnitude (from 10^{19} GeV for the Planck scale to 246 GeV for the Fermi scale), and it is this huge discrepancy in energy scales that explains why gravity is so much weaker. To produce standard light, we can simply accelerate electrons in a bulb. By contrast, gravity's strength is so weak that to produce glight of similar amplitude, astronomical objects such as stars or black holes must accelerate and merge in cataclysmic events. Nowadays, every physics undergraduate student has the opportunity to experiment with the quantum nature of light as part of their standard laboratory work. Sadly, equipping undergraduate laboratories with neutron stars may run a little pricey (not to mention the problems presented by the safety risk assessment forms), so probing the quantum nature of gravity is unlikely to become part of the standard undergraduate curriculum anytime soon. Many of the greatest minds of the past century have pondered how we might be able to detect individual gravitons,

5. This is also why there is so much interest in exploring particle physics through particle colliders at around those energy scales, as is currently taking place at the Large Hadron Collider at CERN (Conseil Européen pour la Recherche Nucléaire).

with no definitive answer to date. Of the thought experiments that have been proposed, Freeman Dyson's 2012 Poincaré Prize Lecture contains some of the most well-developed ideas [23].

The gravitational waves we can currently detect on Earth have a typical amplitude of 10^{-20}, meaning that two mirrors located 4 km apart in a vacuum tunnel would be displaced by a distance of 4×10^{-20} km, or 10^{-17} m, a hundredth of the size of a proton. At a frequency of about 100 Hz, the waves detected by LIGO carried at least an order of 10^{40} gravitons. That's a lot more than the flux of photons we receive from the Sun! In this case, though, "more" is not synonymous with "better"—not by a long stretch. Receiving a flux of so many gravitons makes it all the more challenging to isolate a single one from the crowd. If 10^{40} gravitons are able to displace two mirrors by a distance that is a hundredth the size of a proton, then this means that, at best, a single graviton at the same frequency would have an amplitude forty orders of magnitude weaker. The mirrors located 4 km apart would be displaced by a distance of 10^{-57} m, well below the fundamental Planck distance scale. Achieving such a level of precision is not merely technologically difficult, it is simply impossible because it would violate Heisenberg's uncertainty principle.

This suggests that detecting a single graviton with LIGO, Virgo, KAGRA, or any other gravitational wave interferometer located on Earth is likely impossible. How about sending instruments into space? If we keep an open mind, we can imagine a gravitational wave interferometer with arms as long as the Universe itself (remember I am a theorist). Unfortunately, this wouldn't help either. An interferometer is most sensitive to glight with wavelengths of the same order of magnitude as the length of its arms. As a result, the longer the arms of our detector, the wider the wavelength of gravitational waves we will observe, which in turn means a weaker signal and less energy carried by each individual graviton. It appears that it is simply impossible to imagine a gravitational wave interferometer, large or small, where the effect of a single graviton could ever be greater than the fundamental Planck distance scale.

Our final journey is certainly off to a bad start. Perhaps we took a wrong turn? Could there be other ways? As first recognized by Einstein, the quantum nature of light can be more easily proven by observing the photoelectric effect, using atoms as our detectors. Electrons orbiting the nucleus of atoms must commit to a specific quantum orbit and can only move up a level by absorption of a specific quanta of energy—for instance, by absorbing a photon with that specific energy. When shining light on an atom, if the frequency of light is too low, the electrons will never be able to capture those quanta of light and jump up, no matter the intensity of light (i.e., no matter how many photons there are). Just as you need the right key to open a door, you need the right frequency to quicken the electron onto another orbit.

In theory, we could devise a similar mechanism with glight. Shining glight at an atom, "all" we have to show is that electrons are only able to jump up a level when glight of the right frequency is used. This is simple in theory, but producing sufficient glight requires astronomical events—like the merger of two stars or black holes—not something we could attempt in a laboratory. Nor does it seem feasible to take your precious atom within the vicinity of a major astronomical collision to see if the electrons fancy jumping up. Perhaps, instead, you could stay home and use glight from the Sun because collisions of electrons and ions in the Sun's plasma produce a spectrum of thermal glight radiation. Unfortunately, this flow of glight comes along with neutrinos and light, which would completely swamp the signal. You could try shielding the atom or instrument from neutrino and light radiation, only allowing glight to pass through your protective walls. This may be theoretically possible, but with the current known forms of matter the protective walls would have to be so massive that the experiment would collapse into a black hole under its own gravitational attraction, yet another occupational hazard in our perilous journey. Unfortunately, as indicated by Dyson, these are the rather unpleasant side effects of the reality of the Universe in which we happen to live. While they do not prove that detecting the effect of a single graviton is impossible, they do show that it will be astronomically difficult. Perhaps in the coming centuries or millennia our understanding and mastery of dark matter or other sectors of matter will open new possibilities. They might be able to shine glight on our atoms without overwhelming us with light and other more ordinary particles. For the time being, however, these possibilities remain entirely in the realm of science fiction.

We might wonder if there is anything left to try in our quest for the graviton. The answer, in all its simplicity and naivety, may be to use the whole sky as our probe. This may sound even more absurd, but what better to probe the structure of spacetime at the quantum level than the one thing we know with near certainty emerged from quantum gravity at the Big Bang: our entire Universe.

Tiny quantum fluctuations in the curvature of spacetime are believed to occur continuously everywhere throughout the Universe. These fluctuations are so small and insignificant that we have no hope of ever observing them, at least not as they bubble in and out of existence in front of us. However, as we discussed in chapter 4, when the Universe was celebrating its $10^{-33\text{rd}}$ second, most cosmologists believe that it went through a period of inflation or extremely fast, accelerated expansion.[6] During this

---

6. Alternatives to inflation have been proposed, and it is possible that other mechanisms are responsible for stretching out the small quantum fluctuations of curvature to cosmological scales, but all these models agree on the quantum nature

ultrarapid expansion, the quantum fluctuations created were almost instantaneously stretched to cosmological size. The primordial accelerated cosmic expansion, acting as a magnifying glass, stretched the smallest quantum fluctuations to cosmological scales. This may sound entirely crazy until you look up and observe those quantum fluctuations in the sky. We now have confirmation that the light that reaches us from the period of last scattering has been imprinted by these fluctuations.

These observations allow us to probe the quantum nature of the primordial curvature fluctuations. However, this only proves that, at that time, our spacetime was hosting a particle of quantum nature, just like the quantum nature of the electron indicated the *need* for gravity to be quantum but did not *prove* it. To prove that spacetime or gravity itself is quantum in nature, we would need to prove that at the beginning of the Universe, small quantum gravitational fluctuations were also bubbling in and out of existence. These gravitational or tensor fluctuations would then also have been stretched to cosmological scales and have an imprint on the light that reaches from the period of last scattering. Since glight can come in a mixture of two polarizations, it should have specific polarized imprints on light.

Multiple instruments are already looking for them. Some rely on polarimeters (instruments designed to determine the polarization of light), which are boarded on balloons that are launched from Antarctica and drift through the upper layers of the atmosphere, enjoying an unobstructed view of a large portion of the sky. Others are telescopes placed in remote locations like the South Pole or the Atacama Desert in Chile. All

---

and origin of those fluctuations. In alternatives to inflation, there may not be any detectable tensor fluctuations, and it may hence be impossible to prove the quantum nature of gravity in the exact same way.

these instruments are specifically designed to be sensitive to the polarization of the primordial light that reaches us from the cosmic microwave background. Detecting a nontrivial polarization pattern with a specific statistical distribution could be the sign that the sky is imprinted by quantum fluctuations of gravity during the very early Universe, and that gravity is indeed a quantum phenomenon at its most fundamental level.

So far, no experiment has been able to successfully confirm such observations, but the hunt is just getting started. Over the next decade or two, Earth-based and possibly space-based experiments will be able to probe the polarization of the light coming to us from the period of last scattering with increasing accuracy over a large portion of the sky, potentially confirming that it has indeed been imprinted by quantum fluctuations of gravity.

Yet even if the imprints of primordial gravitational waves are not observed on the polarization of light, not all will be lost. Primordial glight has been traveling through the Universe since its production some 13.8 billion years ago. In the near future, it may even be possible to catch some of it. Thanks to future space-based missions like LISA, and missions like the PTA, it may be possible to observe the glow of the oldest glimpse of glight, produced by quantum fluctuations in the early Universe and magnified to observable scales by the Universe itself. This glow of primordial glight would be coming from every direction in the sky, with sufficiently low frequency and precise statistics to indicate that it was generated in the very early stages of our Universe.

The detections of such primordial glight with a spectrum that matches that dictated by quantum fluctuations would not only open a new window into the state of our Universe as it was in its infancy, but it would also serve as another indication that gravity and the very fabric of spacetime are fundamentally quantum in nature, paving the way toward proving the existence of the graviton. Pushing our observations in those directions may eventually reveal whether or not the graviton is massive. It may also provide further hints about the meaning of space and time and potentially about the origin of the Universe.

So the next time you look up at the sky, whether day or night, consider that it is likely imprinted with quantum glight footprints. A message written over 13.8 billion years ago is just above our heads, hiding some of the deepest secrets of our Universe and waiting for us to decode it.

Creatures of Gravity

Gravity is all around us, yet we each come to experience and appreciate it in our own way. For some, the omnipresence of gravity is most apparent when leaning over the edge of a high cliff. For others, the majesty of gravity comes to the fore when observing the silent and gracious twirl of the planets and stars in the dark sky. Others will appreciate how tenacious gravity can be after witnessing a shuttle launch and the immense force it takes to extract ourselves from the Earth's pull. For me, the pursuit of gravity, whether to play with it or simply comprehend it, has shaped my life.

On our adventure together, we have explored the multiple facets of gravity. We have witnessed its metamorphosis from the Newtonian concept of a universal instantaneous force to the Einsteinian understanding of gravity as the curvature of spacetime. That transformation leads to the inevitable tango, into which free-falling dancers form spirals as they travel a straight path through our curved reality. Hidden within the fabric of spacetime, we have discovered the existence of the fundamental force of gravity, one built into the structure of Einstein's theory of general relativity. This gravitational force, while barely noticeable in our everyday lives, manifests itself through tidal effects that have been detected thanks to gravitational wave observatories. Enter a black hole, however, and those tidal forces would both squeeze and tear you apart.

As general relativity celebrates a century of astonishing success, I would be the first to question any model that departed from it, even on large cosmological scales. Yet massive gravity, by providing a concrete alternative, potentially helps us make sense of the late-time acceleration of the Universe and explain why the quantum sea of vacuum energy does not cause the Universe to curl to less than a centimeter in size. As we continue our journey, there is no denying that our quest for a more fundamental understanding of gravity still has a long, if not eternal, way to go. Ultimately, my own involvement will most likely be minuscule, but this is not why we take part in science. Each contribution, even if unsuccessful or doomed from the start, helps us shed light on a small piece of the puzzle. Despite general relativity's unparalleled track record, I would be surprised if we did not make any significant breakthroughs or departures within my lifetime. Time, or gravity itself, will tell.

Diving through the deep layers of gravity, we may come to appreciate how it has shaped who we are and how we evolve, whether the royal "we" refers to the most elementary particle, or past, present, and future human civilizations, or the Universe as whole, or the fabric of reality spanning throughout the multiverse. But as we return from these wonderful journeys, I often wonder whether, pragmatically, we should keep our feet on the ground and our focus on efforts to improve life on Earth right here, right now, rather than trying to solve the mysteries of the great beyond. We should, of course, direct our attention to the urgent matters of our survival and that of our Earth. And yet the progress our species has made over the past thousands of years and the innumerable problems that science has been able to solve are inseparable from the curiosity to explore beyond our front door—to extract ourselves from our most pressing needs and push the frontiers of our knowledge toward the unknown, uncovering the answers to problems we did not even know we had.

When theoretical physicists describe our work and our findings, we are frequently questioned about the practical applicability of our research in a world with so many immediate and significant crises. In the face of such questions, we are perhaps too quick to validate our research by pointing to the impact that previous discoveries have had on real-time consumer-ready deliverables. Far too often, I hear myself arguing that applications from fundamental physics research are always at best very delayed; GPS, I am quick to point out, is only possible because of general relativity, yet that application did not arise for more than half a century after the initial scientific breakthrough. Or I will observe that many of the technologies we develop to advance our research, such as the World Wide Web (invented at CERN), are exported to other sectors of society, where they are put to a variety of beneficial uses. Nowadays, it would simply be impossible to contemplate our lives without the downstream benefits of earlier research—research that, at the time, had no obvious application.

When it comes to contemporary gravitational research, the more open-minded visionary types will ask me if the gravitational research that I do could, for example, help us detect new exoplanets, driving our fundamental quest for extraterrestrial life, as some studies about gravitational waves have suggested. The real question, however, is why stop there? While the discovery of exoplanets and the search for extraterrestrial life are undeniably some of the most exhilarating quests currently underway, a better understanding of gravity may pave the way for discovering new Universes, new dimensions of space, or even new dark sectors of matter and energy. And some of them may be home to entirely new species and civilizations coexisting in parallel right next to us—who knows? Such a discovery would totally redefine not just our place in the Universe but our observable Universe's place in the very structure of reality. To my mind, there is no more profound or fundamental quest. Whatever we as a society hope to achieve, understanding nature at its most fundamental level is the best way to equip ourselves with the tools and techniques to solve the problems of tomorrow.

Yet validating fundamental research through such practical applications—even practical applications that themselves seem utopian—misses an important element of what this type of research is all about. If you've been following our quest for gravity, I don't need to convince you that understanding gravity is not about its direct technological or business utilization nor even about any existing or futuristic application of our research. It is about understanding this remarkable phenomenon that permeates every aspect of our lives and determines the evolution and fate of the whole Universe. We chase gravity because we are creatures of gravity. Whether we walk, dive, fly, or explore space as an astronaut, as we soar and fall, succeed and fail, we all inhabit a Universe defined by gravity—and, as a scientist, I will forever pursue its secrets.

Bibliography

1. "For Mr Bently at the Palace in Worcester A 4th Lett. from Mr Newton [February 25, 1692]." 189.R.4.47, ff. 7–8, Trinity College Library, Cambridge, United Kingdom. In *The Newton Project*, AHRC Newton Papers Project, edited by Rob Iliffe and Scott Mandelbrote, University of Oxford, October 2007. https://www.newtonproject.ox.ac.uk/view/texts/normalized/THEM00258.

2. Maxwell, James Clerk. "VIII. A Dynamical Theory of the Electromagnetic Field." *Philosophical Transactions of the Royal Society of London* 155 (December 1865): 459–512. https://doi.org/10.1098/rstl.1865.0008.

3. Agazie, Gabriella, et al. "The NANOGrav 15 yr Data Set: Observations and Timing of 68 Millisecond Pulsars." *The Astrophysical Journal Letters* 951, no. 1 (June 29, 2023). https://iopscience.iop.org/article/10.3847/2041-8213/acda9a.

4. Einstein, Albert. "Aphorisms for Leo Baeck." In *Ideas and Opinions*, translated by Sonja Bargmann, 27–28. New York: Bonanza, 1954.

5. Einstein, A., B. Podolsky, and N. Rosen. "Can Quantum-Mechanical Description of Physical Reality Be Considered Complete?" *Physical Review* 47, no. 10 (May 1935): 777–780. https://doi.org/10.1103/PhysRev.47.777.

6. Einstein, A., and N. Rosen. "The Particle Problem in the General Theory of Relativity." *Physical Review* 48, no. 1 (July 1935): 73–77. https://doi.org/10.1103/PhysRev.48.73.

7. Weinstein, Galina. "Einstein and Gravitational Waves 1936–1938." arXiv, February 15, 2016. https://doi.org/10.48550/arXiv.1602.04674.

8. Einstein, A., and N. Rosen. "On Gravitational Waves." *Journal of the Franklin Institute* 223, no. 1 (January 1937): 43–54. https://doi.org/10.1016/S0016-0032(37)90583-0.

9. Einstein, Albert. "On a Stationary System with Spherical Symmetry Consisting of Many Gravitating Masses." *Annals of Mathematics*, 2nd ser., 40, no. 4 (October 1939): 922–936. https://doi.org/10.2307/1968902.

10. Penrose, Roger. "Gravitational Collapse and Space-time Singularities." *Physical Review Letters* 14, no. 3 (1965): 57. doi:10.1103/PhysRevLett.14.57.

11. Hawking, Stephen. "The Occurrence of Singularities in Cosmology." *Proceedings of the Royal Society of London A* 294, no. 1439 (1966): 511–521. https://doi.org/10.1098/rspa.1966.0221.

12. Penrose, Roger. "Gravitational Collapse: The Role of General Relativity." *Rivista del Nuovo Cimento* 1 (1969): 252–276. doi:10.1023/A:1016578408204.

13. Hawking, Stephen, and Roger Penrose. "The Singularities of Gravitational Collapse and Cosmology." *Proceedings of the Royal Society of London A* 314, no. 1519 (1970): 529–548. https://doi.org/10.1098/rspa.1970.0021.

14. Lodge, Oliver. "XXXIX. The Density of the Æther." *London, Edinburgh, and Dublin Philosophical Magazine and Journal of Science*, 6th ser., 13, no. 76 (April 1907): 488–506.

15. Kragh, Helge. "Walther Nernst: Grandfather of Dark Energy?" *Astronomy and Geophysics* 53, no. 1 (February 2012): 1.24–1.26. https://doi.org/10.1111/j.1468-4004.2012.53124.x.

16. Lenz, Wilhelm. "Das Gleichgewicht von Materie und Strahlung in Einsteins geschlossener Welt." *Physikalishe Zeitschrift* 27 (1926): 642–645.

17. Enz, Charles P. *No Time to Be Brief: A Scientific Biography of Wolfgang Pauli.* Oxford: Oxford University Press, 2002.

18. Enz, C. P., and A. Thellung. "Nullpunktsenergie und Anordnung nicht vertauschbarer Faktoren im Hamiltonoperator." *Helvetica Physica Acta* 33 (1960): 839–848.

19. Fierz, Markus, and Wolfgang Ernst Pauli. "On Relativistic Wave Equations for Particles of Arbitrary Spin in an Electromagnetic Field." *Proceedings of the Royal Society of London A* 173 (November 1939): 211–232. https://doi.org/10.1098/rspa.1939.0140.

20. de Rham, Claudia, and Gregory Gabadadze. "Selftuned Massive Spin-2." *Physics Letters B* 693 (June 2010): 334–338. https://doi.org/10.1016/j.physletb.2010.08.043.

21. de Rham, Claudia, and Gregory Gabadadze. "Generalization of the Fierz–Pauli Action." *Physical Review D* 82 (2010): 044020. https://10.1103/PhysRevD.82.044020.

22. de Rham, Claudia, Gregory Gabadadze, and Andrew J. Tolley. "Resummation of Massive Gravity." *Physical Review Letters* 106, no. 23 (June 2011): 231101. https://doi.org/10.1103/PhysRevLett.106.231101.

23. Dyson, Freeman. "Is a Graviton Detectable?" Poincaré Prize Lecture, International Congress of Mathematical Physics, Aalborg, Denmark, August 6, 2012. https://publications.ias.edu/sites/default/files/poincare2012.pdf.

Index

Note: Page numbers in *italics* indicate figures.

A Note on the Type

This book has been composed in Arno, an Old-style serif typeface in the classic Venetian tradition, designed by Robert Slimbach at Adobe.

GPSR Authorized Representative: Easy Access System Europe - Mustamäe tee 50, 10621 Tallinn, Estonia, gpsr.requests@easproject.com